PRINCIPLES OF
X-RAY CRYSTALLOGRAPHY

PRINCIPLES OF
X-RAY CRYSTALLOGRAPHY

LI-LING OOI

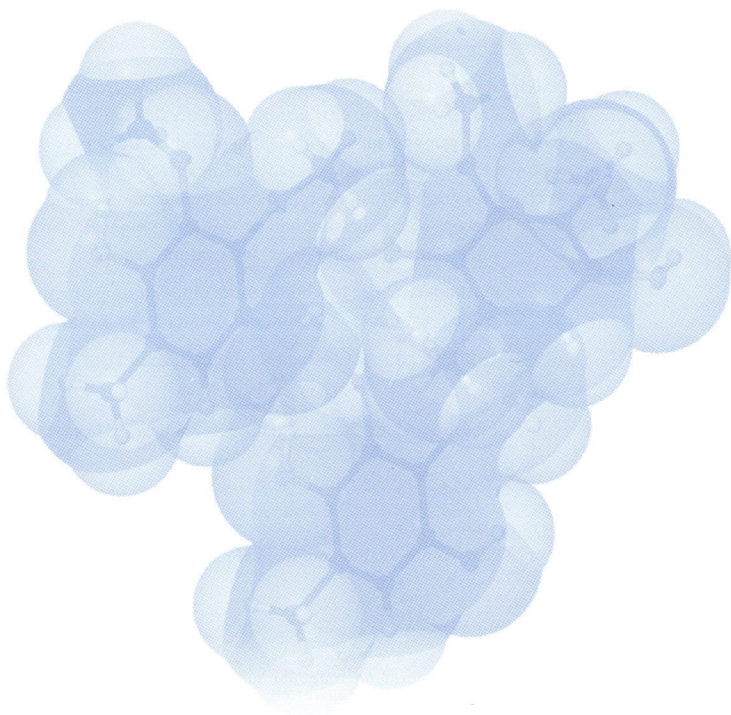

OXFORD

UNIVERSITY PRESS

OXFORD
UNIVERSITY PRESS

Great Clarendon Street, Oxford OX2 6DP

Oxford University Press is a department of the University of Oxford.
It furthers the University's objective of excellence in research, scholarship,
and education by publishing worldwide in

Oxford New York

Auckland Cape Town Dar es Salaam Hong Kong Karachi
Kuala Lumpur Madrid Melbourne Mexico City Nairobi
New Delhi Shanghai Taipei Toronto

With offices in

Argentina Austria Brazil Chile Czech Republic France Greece
Guatemala Hungary Italy Japan Poland Portugal Singapore
South Korea Switzerland Thailand Turkey Ukraine Vietnam

Oxford is a registered trade mark of Oxford University Press
in the UK and in certain other countries

Published in the United States
by Oxford University Press Inc., New York

British Library Cataloguing in Publication Data

Data available

Library of Congress Cataloguing in Publication Data

Ooi, Li-ling.
Principles of x-ray crystallography/Li-ling Ooi.
p. cm.
ISBN 978-0-19-956904-5
1. X-ray crystallography—Textbooks. I.   Title.
QD945.O55   2010
548'.83—dc22

2009039592

Typeset by MPS Limited, A Macmillan Company
Printed in Italy
on acid-free paper by
L.E.G.O S.p.A

ISBN 978-0-19-956904-5

3 5 7 9 10 8 6 4 2

PREFACE

For a long time, X-ray crystallography remained a highly specialized technique that was inaccessible to the average undergraduate or even the postgraduate. However, with the development of more and more sophisticated diffractometers, the steady rise in computing power, and the advances in crystallographic computer software, X-ray crystallography is now often a core subject in many University undergraduate courses.

While there are many excellent crystallography texts available, the majority of them are aimed at the specialist crystallographer, and are often highly technical and detailed, and too advanced to be covered within the remit of an undergraduate course.

The main aim of this book is to provide a *gentle* introduction to X-ray crystallography for the undergraduate or the beginner postgraduate student. To do this, it has been necessary to convey the ideas and principles in ways that may seem abbreviated and simplified – the aim is not to pare down knowledge but to enable an easier grasp of the subject. Further references have been provided so that the engaged and interested student may go on to explore the relevant topics further.

The first five chapters of the book provide an introduction to the foundation and background of X-ray crystallography, starting first with the concepts of crystal systems, Bravais lattices, and unit cells, and moving on to indices, symmetry elements, space groups, and systematic absences. The following two chapters discuss solutions to the phase problem and the refinement of crystal structures.

Chapter 8 describes the crystallographic experiment and Chapter 9 describes the validation of crystal structures and the key parameters to consider when judging a crystal structure – this may be useful both for the trainee crystallographer and for those wishing to learn more about the crystallographic parameters often quoted in scientific publications.

USING THE BOOK

The teaching and learning of X-ray crystallography is undoubtedly challenging and the ability to 'see' in three dimensions is extremely helpful. To help with this, I have included learning tips, learning activities, and self-test questions, which will hopefully be beneficial to a wide range of learning styles.

There are a number of worked examples, which illustrate in a step-by-step manner the process of solving some mathematical-based questions. Interesting points, facts, and reminders are added in boxes at key points within the text. These learning aids are complemented further by resources available online: links to other useful interactive learning tools, step-by-step guides, and practice datasets. Go to **http://www.oxfordtextbooks.co.uk/orc/ooi/** for access to these online materials.

I hope that this book will help to dispel misconceptions that crystallography is a 'difficult' subject and that more students will find fascination and excitement in X-ray crystallography as a result.

ACKNOWLEDGEMENTS

I am most grateful to Jonathan Crowe, Andy Jircitano, Louise Male, Paul Raithby and Georgina Rosair for invaluable advice, insightful comments, and input in the preparation of the text; any errors that remain are completely mine alone.

I am also grateful to Angus Kirkland and Lars Liljas for reviewing the early drafts and providing useful feedback.

Many thanks and much appreciation also to Paul Raithby and Mary Mahon for their unfailing enthusiasm and passionate teaching of crystallography.

Many thanks are also due to Catherine Housecroft for invaluable advice and encouragement.

Also to Emma Lonie and all the team at OUP, many thanks for the excellent work.

Finally I would also like to thank my family and friends for their patience, support, and encouragement.

Acknowledgements are also due to Len Barbour for the use of Crystallography True Type Font.

CONTENTS

In the autumn of 1912, an alarming article appeared in the *Daily Mail* that quickly made its way by special cable across the Atlantic, to the front page of *The New York Times*.[1] 'Molecules rendered visible and test tubes made useless. . . ' This great discovery, which was about to render traditional chemical science research obsolete, was none other than X-ray crystallography.

AN INTRODUCTION TO THE METHOD

1

By the end of this chapter you should be able to:

- Gain an appreciation of crystallography as a technique;
- Understand the concept of the unit cell, and of the asymmetric unit in a crystal lattice;
- Recognize and identify the 7 crystal systems and the 14 Bravais lattices.

Contrary to predictions made almost a century ago, X-ray crystallography as a technique has certainly not caused the premature death of chemical research; instead it has greatly influenced and contributed to our knowledge and understanding of the world around us, and will doubtless continue to do so.

Have you ever wondered how we know today that a humble crystal of table salt consists of sodium and chloride ions arranged in a cubic close-packed structure [2] or how is it known that graphite and diamonds [3] are both simply carbon atoms arranged differently? Or perhaps you might have questioned how Watson, Crick, and Wilkins determined that the basic building block of life, DNA (deoxyribonucleic acid), consists of a double helix structure.

The answers to these questions lie in X-ray crystallography, an analytical technique that uses X-rays to identify the arrangement of atoms, molecules, or ions within a crystalline solid.

1.1 X-RAY CRYSTALLOGRAPHY AS AN ANALYTICAL TOOL

Section learning outcomes

To be able to:

- Distinguish between spectroscopy and diffraction;
- Recognize the differences between small molecule and protein crystallography.

In the fields of chemistry and biology there lies an inherent need to identify, both qualitatively and quantitatively, the components (molecules, atoms, or ions) within any given *matter*. This identification serves not only to inform and educate, but also allows some form of further design or invention to take place once identification has been achieved.

Apart from X-ray crystallography, other analytical methods based on spectroscopy also provide an insight into the content and components of a sample under study.

1.1.1 Diffraction vs spectroscopy

The science of X-ray crystallography is based on the diffraction of X-rays by a crystalline material. It is the only analytical technique that can provide, with uncompromising certainty, the molecular structure of a given compound in the solid crystalline state.

Diffraction can be explained using an analogy to light being refracted by microscope lenses, when a sample is examined under a microscope.

If we refer to Fig. 1.1, (a) shows in outline how a microscope works while (b) is an outline of an X-ray diffraction experiment. A microscope is used when a sample is too small to be seen with the naked eye. When such a sample is examined under a microscope, as seen in Fig 1.1(a), light (from the light source) passes through the sample and is scattered. When the scattered light reaches the microscope lenses, the light is *refracted* and refocused by the lenses. The image of the sample can then be seen by eye through the lens (or eyepiece).

Just as a microscope sample is too small to be seen with the naked eye, the content of a crystal lattice is also too small to be viewed without instrumentation. A typical X-ray diffraction experiment requires only a small single crystal sample, usually of a few micrometres. The crystal sample physically interrupts the flow of X-rays from a source, causing the X-rays to scatter. This form of scattering is known as *diffraction*. The diffracted X-rays are detected

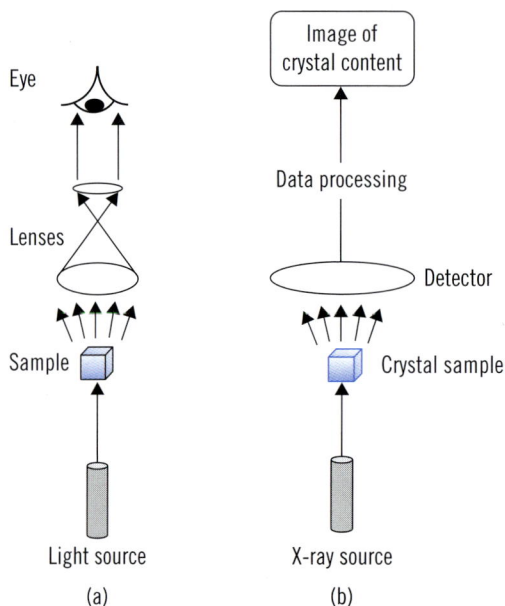

FIGURE 1.1 (a) Refraction of light from a microscope; (b) diffraction of X-rays in a crystallography experiment

by a detector. The way in which X-rays are diffracted by a particular crystal depends on the structure of the crystal, such that every crystal generates a unique diffraction pattern.

Although a detector is comparable to a microscope lens, while a microscope lens is able to refocus light in order to produce an image of the sample, a detector cannot. After the X-ray diffraction is collected by the detector, a range of corrections and computation takes place on a computer. This computer-based process is known as data processing. This process culminates with the generation of a visual image of the content (atoms, molecules, or ions) of the crystal lattice.

Data obtained from an X-ray diffraction experiment allow the direct inference of information. For example, the actual contents of the crystal sample, the types and arrangements of atoms, molecules, or ions within the crystal, and information on the bond lengths and angles can be obtained from an X-ray diffraction experiment. (Further details of the X-ray diffraction experiment are outlined in Section 8.5.)

In comparison, analytical methods based on spectroscopy that are used to identify molecular structures frequently exploit the inherent vibrations within a molecule. When a molecule is irradiated with a low-energy source, such as ultraviolet-visible or infrared radiation, the resulting absorptions or emissions of radiation can be measured and subsequently analysed. Spectroscopic methods, such as infrared spectroscopy, nuclear magnetic resonance (NMR), and Raman spectroscopy, are able to provide information about the energy levels within a molecular system, from which inferences about the molecular connectivity can then be made. These inferences sometimes can be extended to provide geometric information such as bond length and angles.

Examples of information obtained from spectroscopic methods include carbonyl ($-C=O$) and hydroxyl (–OH) stretches in infrared, which allow the presence of these groups to be determined, and the identification of different proton and carbon types from NMR.

While spectroscopy and X-ray diffraction are both very different techniques, in many ways they are complementary. The use of spectroscopy often provides very quick results, whereas an X-ray diffraction experiment can sometimes take days or weeks. The samples for a spectroscopic analysis are also not required to be a crystalline solid and spectroscopy can often be used as a bulk technique. This means that it is often safe to assume that the spectroscopic analysis of a small sample is representative of the bulk.

In contrast, X-ray diffraction provides very detailed, accurate information on the particular single crystal under study. It is necessary for that sample to be a crystalline solid. The results from a single crystal diffraction experiment are often assumed to be representative of the bulk of the sample, but this is not necessarily always true and bulk analysis usually needs to be carried out and reconfirmed using a spectroscopic technique.

While the field of X-ray crystallography relies on the analysis of X-ray diffraction from crystalline samples of very small sizes, in practice, X-ray crystallography can be further subdivided into small-molecule crystallography and macromolecular crystallography.

> *Refraction* refers to a change in the direction of a source of light, when the light moves from one medium to another while *diffraction* is a phenomenon that is often described as 'waves bending around corners'. Diffraction occurs when waves encounter an obstacle.

1.1.2 Small-molecule crystallography

Small-molecule crystallography, in which molecular sizes range from a few atoms to several hundred, is also commonly known as chemical crystallography. This can include inorganic, organic, and organometallic molecules.

Small-molecule crystallography is widely employed in the identification and structural confirmation of newly synthesized molecules in various fields ranging from catalysts and new materials to new drugs. Owing to the small size of the molecules under study in small-molecule crystallography, it is usually possible to identify accurately all of the atom types within the molecule.

Within the crystal lattice of a 'small molecule', it is possible to find a range of intramolecular and intermolecular interactions; hydrogen bonding and π-stacking are examples of solid-state interactions that can be identified within a crystalline solid. These are discussed in further detail in Section 7.4.

1.1.3 Macromolecular (protein) crystallography

In contrast to chemical crystallography, macromolecular crystallography, which is also known as protein crystallography, is used to elucidate the three-dimensional structure of large biological molecules. In general, biological macromolecules can be divided into two groups: nucleic acids and proteins.

Possibly the most famous of all macromolecular structures is that of a nucleic acid, DNA (deoxyribonucleic acid), the elucidation of which won Watson, Crick, and Wilkins the Nobel Prize in 1962. Biologically, nucleic acids store and transmit genetic information. However, only a small number of these nucleic acids can be crystallized for study by X-ray crystallography. These consist mainly of derivatives of DNA, RNA (ribonucleic acid), and various combinations of both DNA and RNA.

Proteins are long-chain polymers (polypeptides) built from a combination of the 20 naturally occurring amino acids. While proteins as a group can be further divided into two subcategories, globular and fibrous proteins, crystallographers are interested mainly in globular proteins as these can be crystallized as single crystals.

The 'globular' nature of these proteins allows a large number of both intramolecular and intermolecular interactions to operate, in particular hydrogen bonding. These interactions both stabilize the protein, and give it its characteristic three-dimensional shape, which contributes to its crystallinity. The medium or solution in which the protein is crystallized also contributes to the stabilization of the crystal structure as these (usually) smaller solvent molecules fill in the spaces of the protein, and are usually extensively hydrogen bonded.

While small-molecule crystallography focuses on the identification of each atom and the positions of these atoms in three-dimensional space, macromolecular crystallography is usually more concerned with identifying the secondary structures – the shapes and motifs of the overall structure – rather than each individual atom. The two main types of motif typically found in protein structures are helices and β-pleated sheets. These are shown schematically in Fig. 1.2. The identification of secondary structures subsequently leads to the determination of the tertiary or quaternary structure, from which usually emerges the overall three-dimensional structure of a protein. An example of a tertiary structure containing both helices and β-pleated sheets can be seen in Fig. 1.3.

While chemical and protein crystallography differ somewhat in the types of molecule under study, the fundamentals of both (outlined in this book) remain congruous and highly similar.

FIGURE 1.2   Secondary protein structures: (a) helices; (b) β-pleated sheets (images kindly provided by J. Crowe)

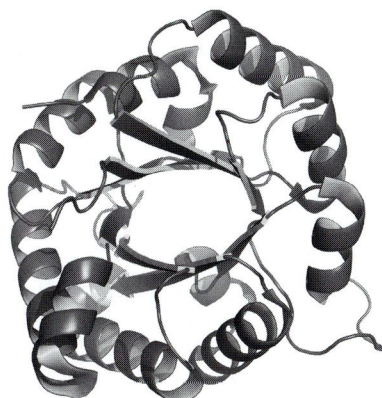

FIGURE 1.3   A tertiary protein structure (image kindly provided by J. Crowe)

SELF-TEST QUESTIONS

1. A newly synthesized chemical compound appears as an amorphous white solid and is soluble in most polar solvents. Which techniques could the chemist attempt in identifying this new compound?

2. Write brief notes comparing the identification processes of spectroscopy and diffraction.

3. What types of protein are suitable for X-ray diffraction and why?

4. Experiment with virtual diffraction online at **http://www.ngsir.netfirms.com/ englishhtm/Diffraction.htm**

1.2   SOLIDS: CRYSTALLINE AND NON-CRYSTALLINE

Section learning outcomes

To be able to:

- Differentiate the different states of matter;
- Distinguish between crystalline and amorphous solids;
- Identify motifs and crystal packing in crystalline solids;
- Recognize two- and three-dimensional repeating patterns.

X-ray crystallography is a solid-state technique, in which samples have to be crystalline solids. X-rays can be diffracted from a crystal due to the periodic arrangement within a crystalline solid. The periodic nature of the crystalline solid is explained in further detail next, while X-ray diffraction is outlined in Section 2.3.

1.2.1   What are solids?

The three primary states of matter are the gaseous, liquid, and solid states. These states are most commonly identified by their physical form but they differ on an atomic or molecular scale too.

Referring to Fig. 1.4(a) we see that the atoms and molecules in solids are ordered and arranged with little movement except for vibrational energy around a central point.

The atoms and molecules within the liquid and gaseous states in Fig. 1.4(b) and (c), however, are less ordered and contain comparably greater amounts of energy. This energy enables the atoms to move. In solids, the movement of atoms is very restrained, akin to vibrations, while in a liquid, the atoms, having more energy, are able to move and take on the shape of the liquid's container. The atoms in gases, however, have the most energy and are able to move the most.

Solids, as we know them in the world around us, can broadly be divided into two distinct groups: *crystalline* and *non-crystalline* (also known as amorphous) solids. As X-ray crystallography relies on the principle of diffraction, it can only be used to analyse crystalline solids.

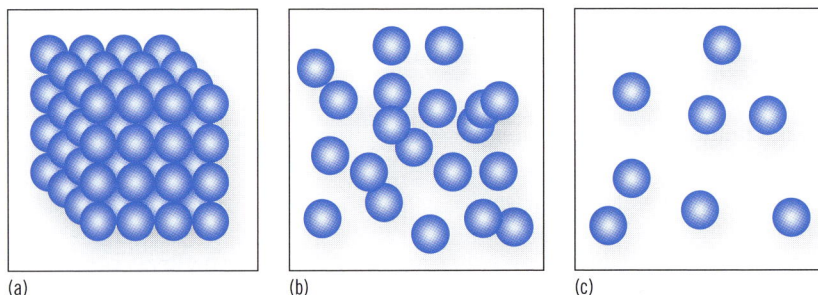

(a)   (b)   (c)

FIGURE 1.4 The atomic representations of the three states of matter: (a) solid; (b) liquid; and (c) gas

1.2.2 Crystalline and amorphous solids

The two main classes of solid, crystalline and amorphous solids (glasses, plastics, semiconductors, for example), are difficult to distinguish physically or visually, as differences lie at the atomic scale within the molecular make-up of each type of solid. While both contain *short-range order,* determined by the chemical bonds within the molecules, the lack of *long-range order* within the amorphous solids makes them unsuited to crystallographic analysis. The lack of long-range order, by definition, means that they cannot form crystal structures – and it is the structure of ordered crystals that is determined by X-ray crystallography.

Referring to the two-dimensional diagram in Fig. 1.5, both types of solid contain hexagonal molecules arranged in two distinctly different ways. In Fig. 1.5(a), we find that the molecules are ordered and arranged with specific distances between specific points or locations and this repetition is translated throughout the crystal. This is known as long-range order or translational periodicity.

Looking at Fig. 1.5(b), however, we notice that while all molecules also contain six atoms, the atoms are not regularly arranged to form uniform hexagonal shapes. While these atoms are not randomly distributed throughout space as in gases, they do lack the repetitive order that gives rise to translational periodicity or *long-range* order that we can find within crystals.

1.2.3 Motifs and packing within crystalline solids

As we now know, a crystalline solid consists of atoms and molecules arranged in a consistent fashion, giving rise to *long-range* order. The specific arrangement of these atoms and molecules within the crystal lattice also gives rise to identifiable motifs. The characteristic arrangement of atoms and molecules within the lattice is known as crystal packing. Figure 1.6(a–c) represents crystal packing in two dimensions, while Fig. 1.6 (d) represents packing in three dimensions. Each diagram consists of an 'object' that is repeated throughout the space, thereby ensuring that each is related to the next by translation.

On closer examination of the diagrams, you can identify the repetitive motifs and how the translations occur. Looking at Fig. 1.6(a), we can identify that each flower is an exact copy of the next; for Fig. 1.6(b), similarly, the four-point star objects are repetitions throughout. Although there are similarities between Fig. 1.6(b) and (c) the use of two colours in Fig. 1.6(c) now mean that the 'object' consists of two four-point stars (one dark and one light).

The *translation* of an object occurs when the object is moved up, down, or sideways without being reflected or rotated.

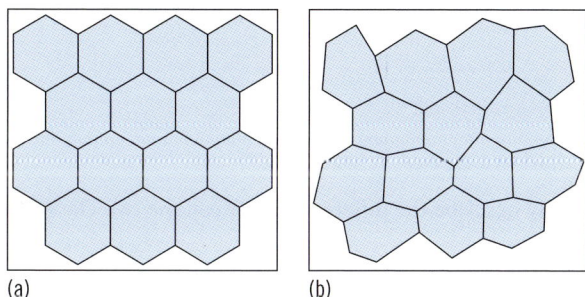

(a)   (b)

FIGURE 1.5 Molecular order within (a) a crystalline solid and (b) an amorphous solid

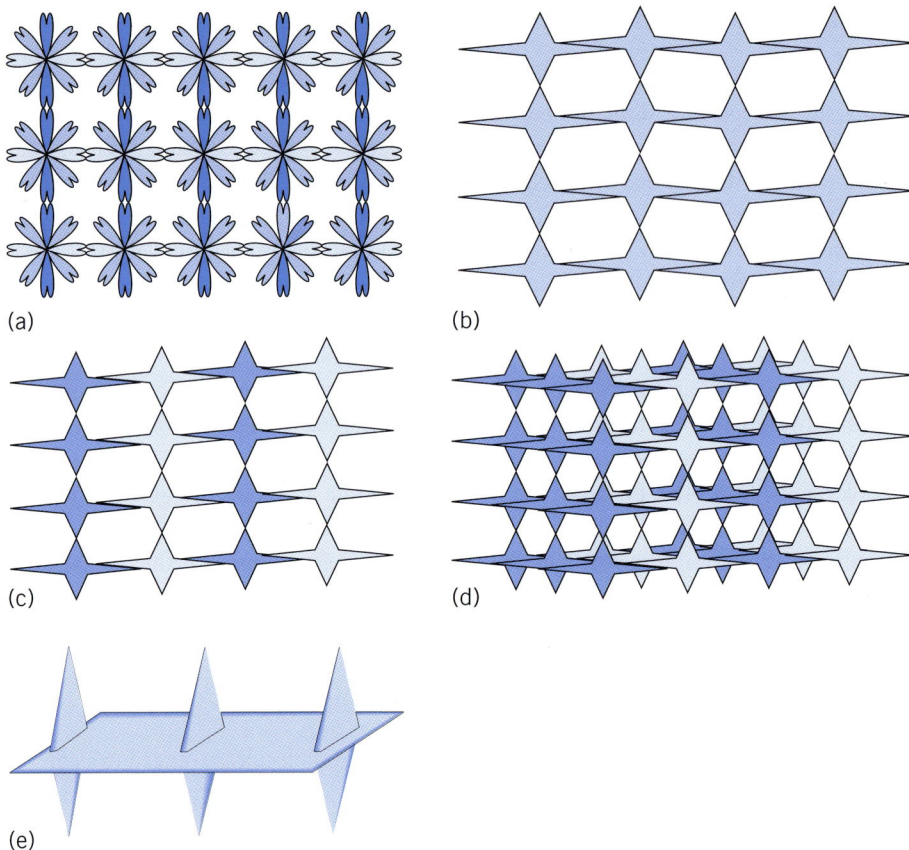

FIGURE 1.6 Examples of two- and three-dimensional packing

The two-dimensional array in Fig. 1.6(c) is repeated in 'depth' in Fig. 1.6(d), giving rise to an example of a three-dimensional packing system. On closer inspection, you will notice that, in three-dimensions, the packing system appears to consist of planes of alternating lines (dark blue and light blue) with peaks and ridges arising at precise intervals. A side-on view (from the left) is shown in Fig. 1.6(e).

A molecular example is given in Fig. 1.7 for coenzyme Q_0 (2,3-dimethoxy-5-methylbenzene-1,4-diol) where Fig. 1.7(a) and (b) represent a single molecule and Fig. 1.7(c) shows a view of part of the three-dimensional crystal lattice.

SELF-TEST QUESTIONS

1. Solids can broadly be divided into two classes. What are they?

2. With the help of diagrams, briefly describe the differences between the two classes of solid.

3. Give an example of *short-range* order and explain how it differs from *long-range* order.

(a)

(b)

(c)

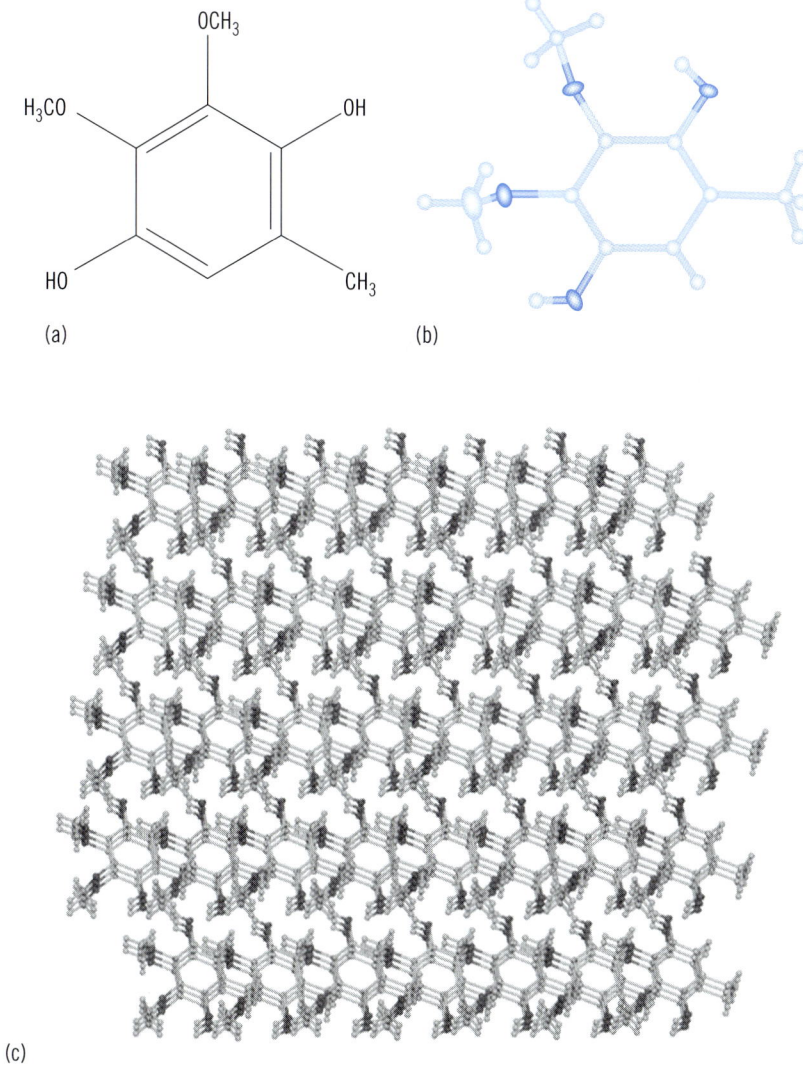

FIGURE 1.7 Coenzyme Q_0: (a) and (b) the molecular structure; (c) the packing diagram

1.3 THE CRYSTAL LATTICE

Section learning outcomes

To be able to:

- Define and identify lattice points, lattice planes, and unit cells;
- Understand and apply the rules for selecting lattice points.

The external morphologies or shapes of crystals can range from cubic or tabular to hexagonal shapes; some examples are shown in Fig. 1.8. However, internally, as we have seen above,

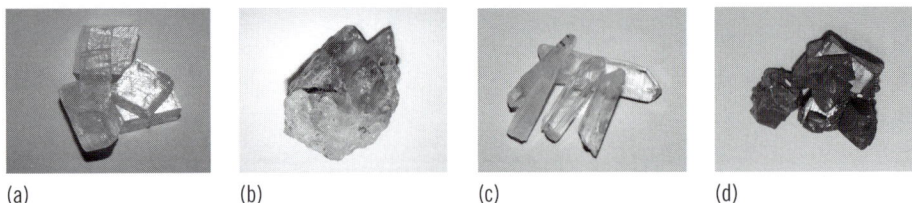

FIGURE 1.8 Different mineral crystals and their crystal morphologies ((a) calcite, (b) amethyst, (c) quartz, (d) pyrite)

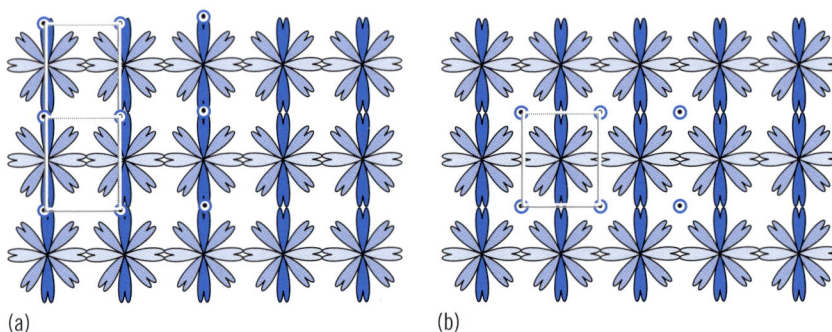

FIGURE 1.9 Examples of different possible lattice points

the atoms and molecules in a crystal form distinctive, organized three-dimensional arrays. We call these organized arrays **crystal lattices**. Specifically, a crystal lattice is defined as an ordered three-dimensional arrangement of ions, atoms or molecules within a solid.

The highly repetitive order within the crystal lattice makes it is possible to identify positions within the lattice that are identical. These positions can then be marked with an imaginary point, referred to as a *lattice point*.

1.3.1 Rules for selecting lattice points

The key to selecting a lattice point is that the environment surrounding each point has to be the same. That is, the view from each lattice point is the same as that from every other point. In two dimensions, for example in Fig. 1.9(a), lattice points are selected at the tips of the dark petals and the surrounding environment of each lattice point consists primarily of the dark petals and the space between the 'flowers'. In Fig. 1.9(b) however, the lattice points are defined in the middle of the space between the 'flowers' and the immediate surrounding environment of each lattice point is mainly the white space and the blue petals.

It is important to point out that lattice points do not represent atoms or molecules, although sometimes they may be located on a particular atom within the crystal lattice. To understand this better, use the learning tip on visualization. (Refer also to Section 4.4).

Learning tip **(visualization) If you were to imagine yourself standing on a lattice point and looking around, then the view all around should be exactly the same as that from any other lattice point.**

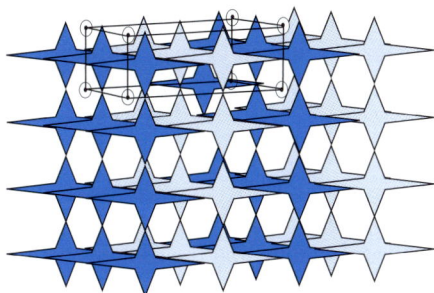

FIGURE 1.10 Identifying a unit cell

1.3.2 Lattice points and lattice planes

In two dimensions, we can then join four adjoining or adjacent lattice points to form a *lattice plane*. In Fig. 1.9(a) for example, it is possible to place a lattice point at the tip of the dark petals of each 'flower' and subsequently join the four lattice points to form a lattice plane that consists of two adjacent halves of the 'flower'. In Fig. 1.9(b), however, the lattice points are connected to form a lattice plane that contains a whole 'flower'.

When considering objects in three dimensions, the adjoining lattice points that are selected can be connected to form a 'box' or a container as illustrated in Fig. 1.10. The box is known as a *unit cell*. The unit cell can then be translated along (up or down, or sideways) through the lattice and its contents remain the same.

SELF-TEST QUESTIONS

1. Identify the possible lattice points in each of the diagrams in Fig. 1.6(a)–(c).

2. Using some samples of wrapping paper, identify some possible lattice points in each sample.

3. (a) Identify the lattice points in some of the symmetry-based pictures by M. C. Escher. These can be found online at **http://www.mcescher.com** under Picture Gallery and Symmetry.

    (b) Discuss how the use of colour may influence the selection of the lattice points.

1.4   THE UNIT CELL AND ASYMMETRIC UNIT

Section learning outcomes

To be able to:

- Understand the association of crystal lattice with unit cell and asymmetric unit;
- Describe and recall the parameters used to define the unit cell.

Once selected, the unit cell forms the basic building block of the crystal lattice akin to the stacking of bricks in the building of a wall. The unit cell can then be reduced further to identify asymmetric units, as explained further in Section 1.4.2.

1.4.1 Definition of unit cell

A unit cell forms the basic building block of a crystal lattice. Each 'box' or container that defines the unit cell is related to the next unit cell by translation and the contents within each unit cell are exactly the same.

For example, if we refer to Fig. 1.11(a), the lattice points are selected so that each unit cell (or box shape) contains one dark blue and one light blue object. The unit cell and its contents are shown in Fig. 1.11(b).

In comparing Fig. 1.12(a) and Fig. 1.12(b), we notice that there are a number of possible choices in selecting the lattice points that make up a unit cell. How then do we decide which is the 'right' unit cell?

If we consider a two-dimensional plane of spots, as shown in Fig. 1.13, there are again several possible ways to select a 'unit cell'; two examples are shown. If we are to compare the two, both are derived from adjacent lattice points.

On closer examination, we will notice that the square has a higher internal symmetry than the diamond shape. We are able to divide the square into a greater number of smaller equivalent sections than the diamond. Hence, the square is a *better* 'unit cell'.

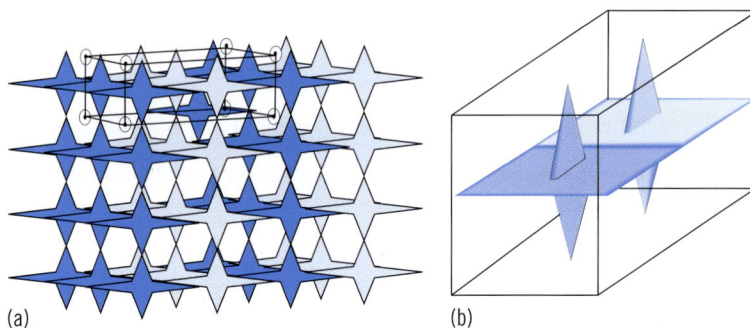

(a)   (b)

FIGURE 1.11   (a) The unit cell identified in and (b) A side-on perspective of the same unit cell

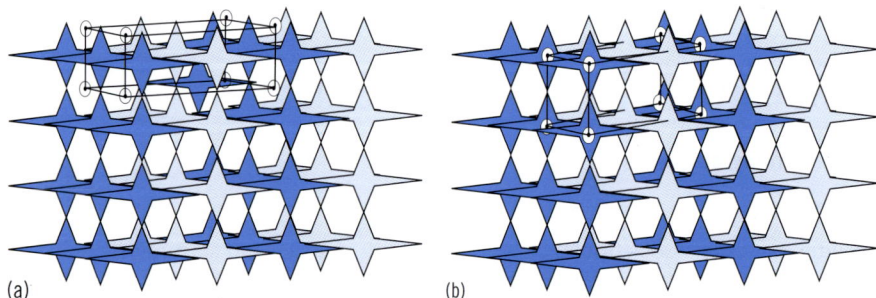

(a)   (b)

FIGURE 1.12   (a) The unit cell as selected in Figure 1.11; (b) an alternative unit cell

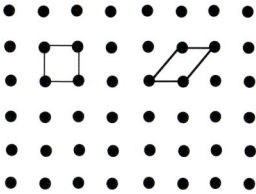

FIGURE 1.13 Two possible 'unit cells'

The same approach is used in three dimensions to select the 'best' unit cell. In summary, the rule for selecting the 'best' unit cell is to select the *smallest unit* that displays the *maximum symmetry* of the structure.

1.4.2 Definition of asymmetric unit

While a unit cell forms the basic building block of crystal, the contents within a unit cell can usually be reduced further to two or more asymmetric units that are related by symmetry. An asymmetric unit can be defined as the *smallest repeat unit* from which the crystal can be constructed. Each asymmetric unit is related by a symmetry element, for example rotation, reflection, or inversion, as detailed in Chapter 3.

Figure 1.14(b) to (d) shows how the asymmetric unit can be derived from the selected unit cell in Fig. 1.14(a). If we are to consider the symmetry in the unit cell in Fig. 1.14(b), we find that there are two planes along which the unit cell is symmetrical; we will label them x and z. If we then divide the unit cell along the vertical plane of z, as shown in (c) and along the horizontal plane x, as shown in (d), we find that the asymmetric unit is a quarter of the unit cell. We can also say that the unit cell consists of four asymmetric units.

> Each asymmetric unit can be related to the next by a symmetry element.

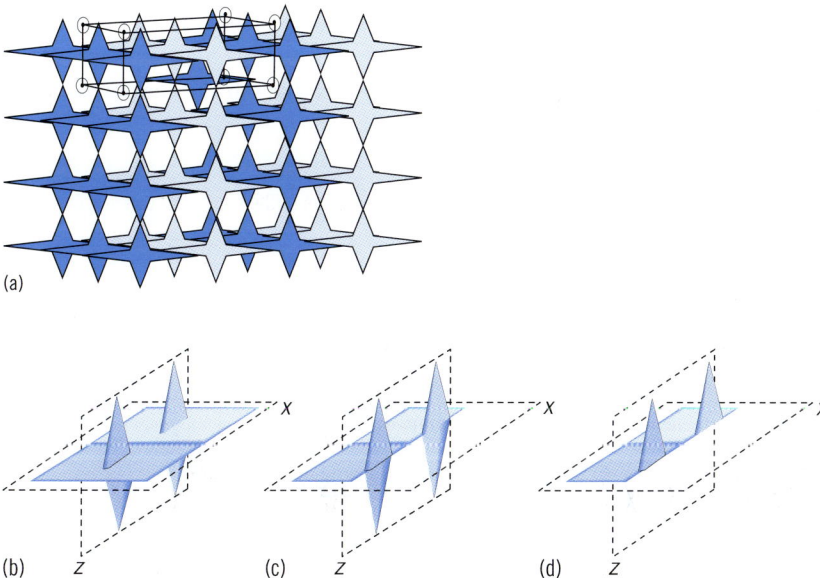

(a)

(b) z (c) z (d) z

FIGURE 1.14 Deconstructing a unit cell to asymmetric units

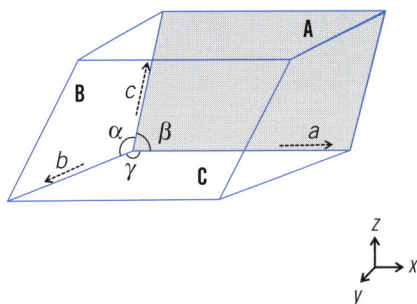

FIGURE 1.15 The parameters of a unit cell

A unit cell usually contains two or more asymmetric units, with only one exception (the exception of the triclinic space group P1 in which the asymmetric unit is the same as the unit cell – refer to Chapter 5). Each asymmetric unit is related to the next by a form of symmetry, which is detailed in Chapter 3.

1.4.3 Unit cell nomenclature and parameters

Once a unit cell has been selected, parameters are used to identify the unit cell's axes and angles. These parameters provide a standard labelling scheme for unit cells.

Referring to Fig. 1.15:

The axes: if the a-axis is assumed to be the horizontal axis and the c-axis the vertical axis, then the b-axis is the axis that is perpendicular to both. If you assume the bottom edge of this book to be equivalent to the a-axis and the spine of the book to be the equivalent of the c-axis, then the b-axis is coming directly out of the book towards you.

The angles: the α angle lies between the b and c axes while the β angle lies between a and c. The γ angle lies between the a and b axes.

The faces: the A faces lie perpendicular to the a-axis (the front and the back of the unit cell) and the B faces are perpendicular to the b-axis (the left and the right side faces of the unit cell) while the C faces are perpendicular to the c-axis (the top and the bottom faces of the unit cell.

1.5 LATTICE TYPES

Section learning outcomes

To be able to:

- Recognize and identify the 7 crystal systems and the 14 Bravais lattices;
- Explain lattice reduction with the aid of diagrams.

Although it may seem that there is an infinite array of possibilities in selecting a unit cell, in reality there are only seven possible 'box' shapes that can be stacked together within a crystal lattice. These are known as the crystal systems. These seven crystal systems are cubic, tetragonal, orthorhombic, hexagonal, trigonal, monoclinic, and triclinic.

In some crystal lattices, it is also possible to identify lattice points other than those defining the unit cell.

The *primitive* (P) lattice is a lattice type in which the lattice points lie only at the corners of the unit cell. This type of lattice is denoted by a P and is found in all the crystal systems.

The *body-centred* (I) lattice is a lattice type in which the lattice points lie at the corners of the unit cell and one lattice point lies in the middle of the unit cell. This type of lattice is denoted by an I and is found in the cubic, tetragonal, orthorhombic, and monoclinic crystal systems.

The *face-centred* (F) lattice is a lattice type in which the lattice points lie at the corners of the unit cell and one lattice point lies in the middle of every face of the unit cell. This type of lattice is denoted by an F and is found in the cubic and orthorhombic crystal systems.

The other *single-face-centred (base-centred)* lattice types are those in which the lattice points lie at the corners of the unit cell and a lattice point lies at each of the two relevant faces. For example, the C-face centred lattice consists of lattice points at the corners of the unit cell and in the middle of both C faces of the unit cell. It is also possible for lattice points to be denoted on the A or B face, although these are considered to be equivalent and are not usually considered separately.

All three types of face centring can be found in the orthorhombic crystal systems while only the C-centred lattice can also be found in the monoclinic crystal system.

These additional lattice points, in conjunction with the seven crystal systems, form the 14 Bravais lattices.

1.5.1 The 7 crystal systems and 14 Bravais lattices

The seven possible unit cell shapes (see Table 1.1) and also known as crystal systems, are identified by their unit cell parameters. The unit cell shapes range from a cubic cell, with the highest possible symmetry, to a triclinic unit cell, with the lowest possible symmetry.

Cubic

In a *cubic* unit cell, all axes are of equal length and all angles are 90°. In essence, this cube shape, with no other lattice points, is the primitive lattice type (P). It is also possible within this unit cell type to identify other lattice points: the body-centred lattice point (I); and the face-centred lattice points (F; on all faces). These give rise to the different Bravais lattices for the cubic unit cell.

$a = b = c,$
$\alpha = \beta = \gamma = 90°.$

Tetragonal

In a *tetragonal* unit cell, only the a and b axes are of equal length, with all angles at 90°. This cell type is similar to a rectangular-shaped box. The only two types of Bravais lattice that occur for this cell type are the primitive-type lattice (P), and the body-centred lattice (I).

$a = b \neq c,$
$\alpha = \beta = \gamma = 90°.$

The German physicist, **Moritz Ludwig Frankenheim** (1801–1869) was the first to categorize correctly the crystallographic lattices; however he had noted 15 lattice types. In 1850, **Auguste Bravais** (1811–1863), a French physicist, corrected Frankenheim's list of lattices, which contained an error of a repetition and subsequently the 14 types of crystallographic lattices today bear his name: Bravais lattices. (It is not certain which lattice was repeated in Frankenheim's list.)

TABLE 1.1 The 7 crystal systems and 14 Bravais lattices

Crystal system	Cell parameters	Lattice types
Cubic	$a = b = c,$ $\alpha = \beta = \gamma = 90°$	P I F
Tetragonal	$a = b \neq c,$ $\alpha = \beta = \gamma = 90°$	P I
Orthorhombic	$a \neq b \neq c,$ $\alpha = \beta = \gamma = 90°$	P I F A, B, or C
Hexagonal	$a = b \neq c,$ $\alpha = \beta = 90°;$ $\gamma = 120°$	P
Trigonal	$a = b \neq c,$ $\alpha = \beta = 90°;$ $\gamma = 120°$	P
Monoclinic	$a \neq b \neq c,$ $\alpha = \gamma = 90°; \beta \neq 90°$	P C
Triclinic	$a \neq b \neq c,$ $\alpha \neq \beta \neq \gamma \neq 90°$	P

Orthorhombic

In an *orthorhombic* unit cell, all the axes are different, although all angles are 90°. This again mimics a rectangular box. The types of Bravais lattice that occur for the orthorhombic cell are the primitive-type lattice (P), the body-centred lattice (I), and face-centred lattice points (F), on all faces and centring on a single face, the A, B, or C faces.

$a \neq b \neq c,$
$\alpha = \beta = \gamma = 90°.$

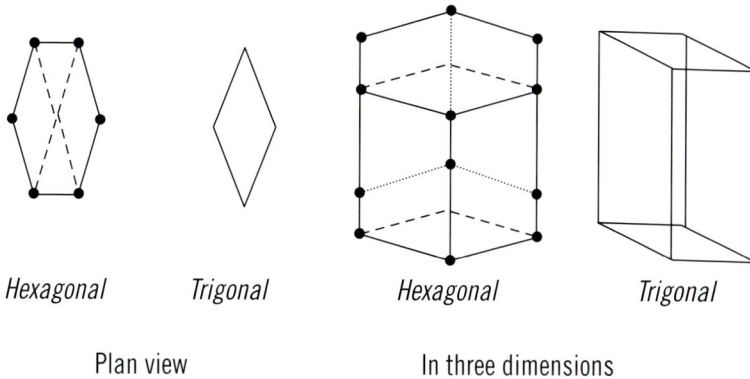

Hexagonal	Trigonal	Hexagonal	Trigonal

Plan view   In three dimensions

FIGURE 1.16 The trigonal and hexagonal unit cells

Hexagonal

In a *hexagonal* unit cell, both the top and bottom faces are hexagonal with the a and b axes of equal lengths and joined by a c-axis of a different length. The angles where the faces meet, α and β, are 90°, while the angles on the hexagonal face are 120°. Only the primitive-type (P) of Bravais lattice is found in the hexagonal unit cell.

$a = b \neq c,$
$\alpha = \beta = 90°; \gamma = 120°.$

Trigonal

The *trigonal* unit cell can be derived from the hexagonal cell. As shown in Fig. 1.16, the hexagonal cell can be subdivided into three trigonal cells (two whole cells and two half cells). Like the hexagonal cell, the axes a and b are of equal lengths and joined by the c-axis of a different length. The angles where the faces meet, α and β are 90°, while the γ angle on the trigonal face is 120°. Again, only the primitive-type (P) of Bravais lattice is found in the trigonal unit cell.

$a = b \neq c,$
$\alpha = \beta = 90°; \gamma = 120°.$

Monoclinic

The *monoclinic* unit cell has all its axes of different lengths with only the angles of α and γ equal to 90°. The primitive (P) and the single-face-centred (C) type of Bravais lattice can be found for the unit cell type. Both the body-centred (I) and the face-centred (F) Bravais lattice types can also occur in the *monoclinic* crystal system; however, as they can also be represented as C-type Bravais lattices, they are usually not considered separately.

$a \neq b \neq c,$
$\alpha = \gamma = 90°; \beta \neq 90°$

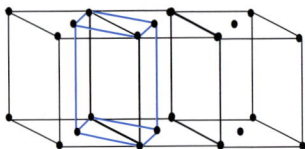

FIGURE 1.17 The face-centred cubic lattice (C) is reduced to the primitive-type tetragonal lattice

Triclinic

Of all the crystal systems, the triclinic unit cell has the least symmetry. The axes are all different and no angle is equivalent to 90°. Only the primitive-type (P) Bravais lattice occurs in the triclinic crystal system.

$$a \neq b \neq c,$$
$$\alpha \neq \beta \neq \gamma \neq 90°$$

1.5.2 Lattice reduction

We have now seen that there are four different types of Bravais lattice, P, I, F, and C. Given that there are seven different crystal systems, we might expect there to be 28 (4×7) different Bravais lattices. In practice, however, not all of the known Bravais lattices occur in all the crystal systems. The reason that there are only 14 Bravais lattices is that lattices can be reduced to a more effective lattice type.

For example, the C-type Bravais lattice does not occur for the cubic unit cell, as shown in Fig. 1.17. If we refer to Fig 1.17, the unit cells drawn in black denote a C-centred cubic lattice. However, this unit cell can be reduced further to a smaller unit cell with higher symmetry and this is shown with the lines in blue. The lines in blue are drawn to join the face-centred lattice points to the corners of the lattices. In doing so, a new lattice type can be found; the primitive-type (P) tetragonal lattice is obtained. This satisfies the rules for selecting the 'best' unit cell.

SELF-TEST QUESTIONS

1. Which crystal system consists of equivalent angles but has axes of different lengths?

2. List the unit cell parameters for a monoclinic crystal system.

3. Examine Fig. 1.12: which unit cell would you select as the 'best' unit cell? Explain the reasons for your choice.

4. Briefly explain how the trigonal and hexagonal unit cells are related.

5. Explain, with the aid of diagrams, why the face-centred (F) lattice type does not occur for the tetragonal crystal system.

◉ CHAPTER SUMMARY

1. X-ray crystallography is a solid-state analytical technique based on diffraction of crystalline solids.

2. Chemical or small-molecule crystallography identifies the detailed atomic contents of a molecule within a crystal lattice

3. Macromolecular or protein crystallography mainly examines globular proteins where the recognition of 'motifs' such as helices and β-sheets are important.

4. Lattice points within a crystal are determined by the environment surrounding that point. All lattice points have the same surrounding environment.

5. A unit cell is a box shape obtained by joining up lattice points in three dimensions.

6. A unit cell, the building block of a crystal lattice, is defined as the smallest unit that displays the maximum symmetry of the structure.

7. The asymmetric unit is the smallest repeat unit within a crystal lattice. The asymmetric units in a unit cell are related by symmetry.

8. A unit cell is defined by its axes, a, b, and c, and its angles, α, β, and γ.

9. There are 7 crystal systems (triclinic, monoclinic, orthorhombic, tetragonal, trigonal, hexagonal, and cubic) and 14 Bravais lattices.

▤ REFERENCES

1. *The New York Times* (13 September, 1912). **http://query.nytimes.com/gst/abstract.html?res=9 A04E0D8123CE633A25750C1A96F9C946396D6CF**.
2. Bragg, W. L. (1914). The structure of some crystals as indicated by their diffraction of X-rays. *Proceedings of the Royal Society* (*London*) **A89**:248–77.
3. Bragg, W. H. and Bragg, W. L. (1913). The structure of the diamond. *Nature* **91**:557.

▥ FURTHER READING

Clegg, W. (1998). *Crystal Structure Determination*. Oxford Primer No 60. Oxford University Press, New York.

Clegg, W., Blake, A. J., Gould, R. O., Main, P. (2001). *Crystal Structure Analysis: Principles and Practise*. IUCr Monographs on Crystallography. Oxford University Press, New York.

Giacovazzo, C., Monaco, H. L., Artioli, G., *et al.* (eds) (2002). *Fundamentals of Crystallography*, 2nd edn. IUCr Texts on Crystallography. Oxford University Press, New York.

Glusker, J. P. and Trueblood, K. N. (1985). *Crystal Structure Analysis – A Primer*. Oxford University Press, New York.

Hammond, C. (2001). *The Basics of Crystallography and Diffraction*. IUCr Texts on Crystallography. Oxford University Press, New York.

LINKS

Online learning tools:

A Java applet to demonstrate diffraction: http://www.ngsir.netfirms.com/englishhtm/Diffraction.htm

A Java applet to demonstrate crystal structure and the 14 Bravais lattices: http://jas.eng.buffalo.edu/education/solid/genUnitCell/index.html.

Here's a bit of fun: A musical introduction to protein structure: http://whozoo.org/mac/Music/Primer/Primer_index.htm

THE CRYSTAL CONSTRUCT

2

By the end of this chapter you should be able to:

- Understand and appreciate the role of Weiss and Miller indices;
- Identify and determine Miller indices and planes;
- Briefly outline diffraction from a crystal lattice;
- Illustrate and explain Bragg's law
- Understand the relationship between real and reciprocal space.

2.1 INTRODUCTION

In Chapter 1, we examined how the contents of a crystal lattice can be classified into the various Bravais lattice types and subsequently reduced to asymmetric units. To gain a better understanding of unit cells and the Bravais lattices in a crystal, it is helpful to be able to visualize the crystal lattice in terms of planes on a familiar three-dimensional plot. As with lattice points and unit cells, the characterization of a crystal lattice into lines and planes is imaginary, and aids our understanding of three-dimensional crystal structure.

In this chapter, we will first examine how the crystal lattice can be further defined by Weiss and Miller indices. We then go on to discuss the phenomenon of X-rays in the context of X-ray diffraction. We will then examine how X-rays are diffracted by crystals and how this can be explained by Bragg's law.

Finally, we consider the resultant diffraction patterns and how they are related to the concept of reciprocal space.

2.2 WEISS INDICES (CELL PARAMETERS), MILLER INDICES, MILLER PLANES

Section learning outcomes

To be able to:

- Explain and identify Weiss and Miller indices;
- Illustrate Miller indices in three dimensions;

- Understand and appreciate the association of Miller indices and unit cell directions with a Cartesian plot.

2.2.1 Weiss indices

Christian Samuel
Weiss (1780–1856) is
recognized for defining
the parameters
in modern
crystallography and
defining the crystal
systems.

Once a unit cell is defined, and the lengths of the cell axes (a, b, and c) determined, these can be placed and mapped against a three-dimensional Cartesian-like plot, where the axes are defined by x, y, and z.

The points at which the unit cell intercepts the axes are also known as Weiss indices. Looking at Fig. 2.1, if we consider an *orthorhombic P-type* cell in which all angles are equivalent to 90°, and all unit cell lengths are not equal, where $a = 2$, $b = 1$, and $c = 3$, then we can also, in considering both the positive ($+$) and negative ($-$) portions of the plot, express the cell parameters as **2a:1b:3c**. These numbers are also known as *Weiss indices*.

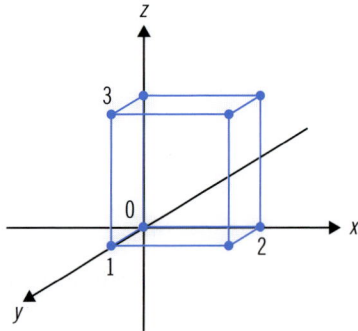

FIGURE 2.1 Weiss indices for an *orthorhombic P* cell

SELF-TEST QUESTIONS

1. Identify the Weiss indices for the following lattices:

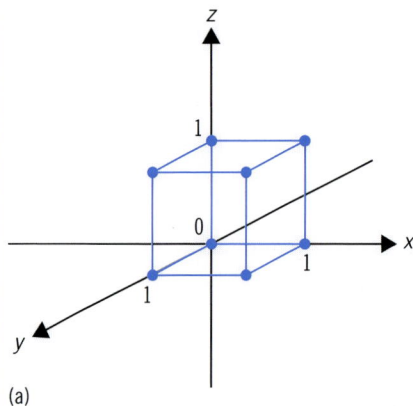

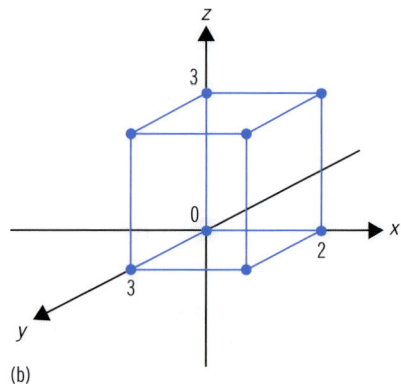

(a)   ;   (b)   ;

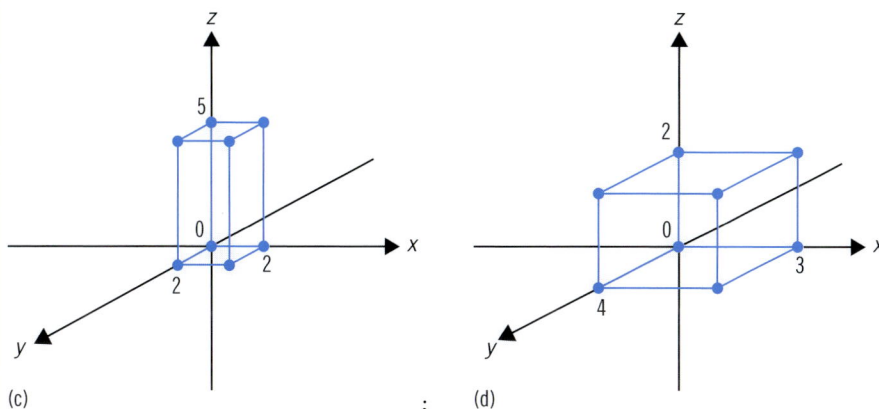

(c)

(d)

2. Draw lattices with the following Weiss indices (assume that all angles are 90°):

(a) $2a:b:2c$;

(b) $3a:b:5c$;

(c) $a:3b:4c$.

2.2.2 General principles for Miller planes and Miller indices

While we now understand and are able, to a certain extent, to visualize the repetitive order of atoms and molecules throughout the crystal lattice, it is still necessary, however, to be able to 'break down' the crystal lattice and subsequently the unit cell further, into imaginary planes.

These planes, known as Miller planes, can be identified by a set of *three* integer values (zero, positive or negative values), h, k, l, known as Miller indices. Miller indices generally refer to sets of parallel planes. The use of Miller indices is also helpful in explaining and further understanding diffraction patterns that result from an X-ray diffraction experiment. This is discussed further in Section 2.3.

The integers corresponding to Miller indices are written in brackets (h, k, l) and are comparable to both the Cartesian directions and the axes of a unit cell, whereby h is commonly equivalent to the *x-axis* or the a direction of the unit cell, k is determined from the y or the b direction and l from the z or the c direction. These are summarized in Table 2.1.

Figure 2.2 shows sets of parallel Miller planes within a crystal lattice. In crystallography, these are also known as Miller sets.

Miller indices were introduced in 1839 by the British mineralogist **William Hallowes Miller** (1801–1880).

TABLE 2.1 Equivalence of Cartesian coordinates, unit-cell dimensions, and Miller indices

Cartesian coordinates	x	y	z
Unit-cell dimensions	a	b	c
Miller indices	h	k	l

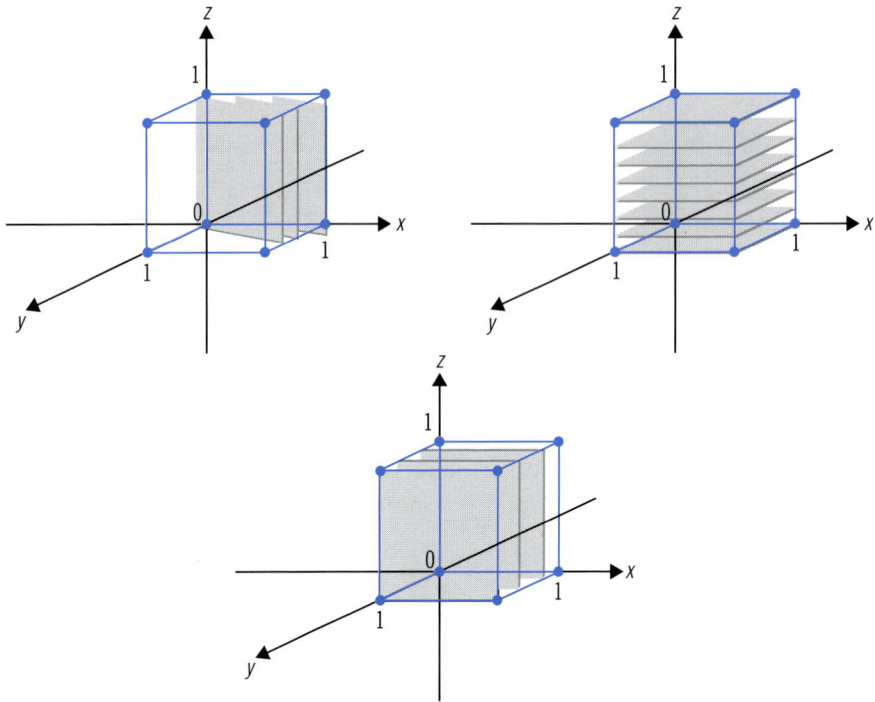

FIGURE 2.2 Multiple Miller planes are also known as Miller sets

2.2.3 Visual information from Miller indices

The integer values of Miller indices can often provide a quick visual representation of the planes in the crystal lattice. For example, if a Miller index is zero, the plane is parallel to that axis. The smaller a Miller index, the more parallel the plane is to the axis, while the larger a Miller index, the more perpendicular the plane is to that axis.

It is important to note that multiplying or dividing a Miller index by a constant has no effect on the orientation of the plane and that Miller indices are almost always small integer values.

2.2.4 Determining Miller indices and identifying Miller planes

Miller indices are determined from the intercepts of the planes along the crystallographic axes. This is done by:

(a) First, *determining the intercepts* of the face along the crystallographic axes (if a plane is parallel to an axis, its value is infinite, ∞);

(b) Then, taking the reciprocals;

(c) Clearing the fractions; and

(d) Finally, reducing the indices to lowest terms.

▷ **WORKED EXAMPLE 2.1**

(a) Determine the Miller index of the shaded area:

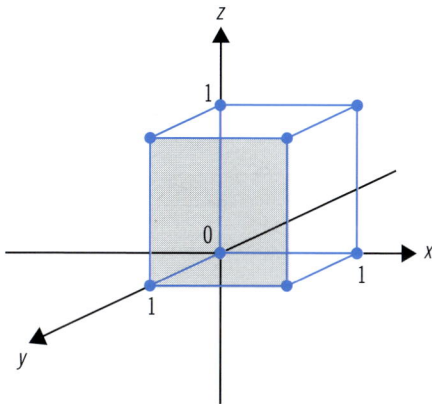

Along	X	y	z
(a) Determine the intercepts	∞	1	∞
(b) Take the reciprocals	1/∞	1/1	1/∞
(c) Clear the fractions	0	1	0
(d) Reduce to lowest terms	0	1	0
Miller index	**(0, 1, 0)**		

(b) Determine the Miller index of the shaded area:

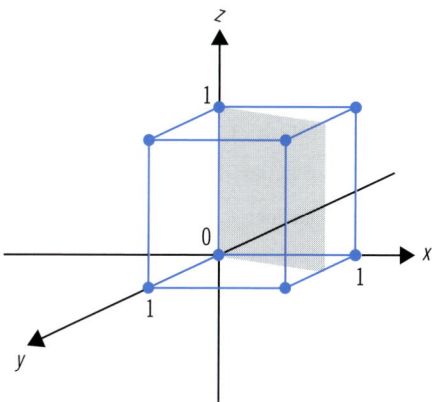

Along	X	y	z
(a) Determine the intercepts	-1	1/2	∞
(b) Take the reciprocals	-1/1	2/1	1/∞
(c) Clear the fractions	-1	2	0
(d) Reduce to lowest terms	-1	2	0
Miller index	**($\bar{1}$, 2, 0)**		

SELF-TEST QUESTIONS

1. Determine the Miller Indices of the shaded areas in the diagrams below.

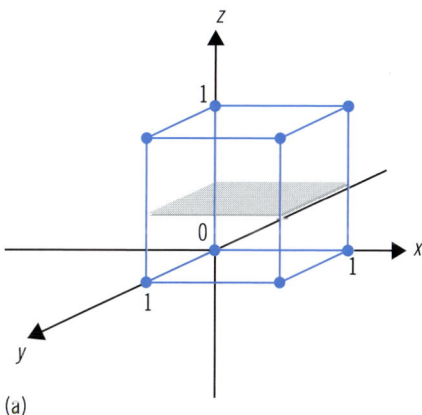

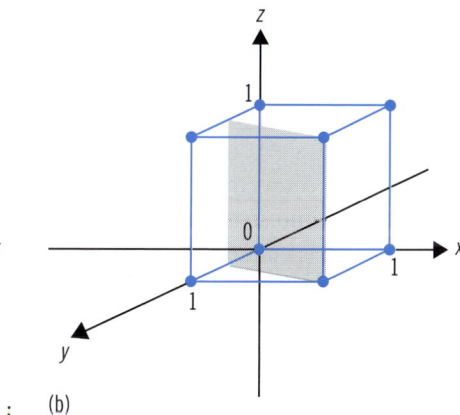

(a) ; (b) ,

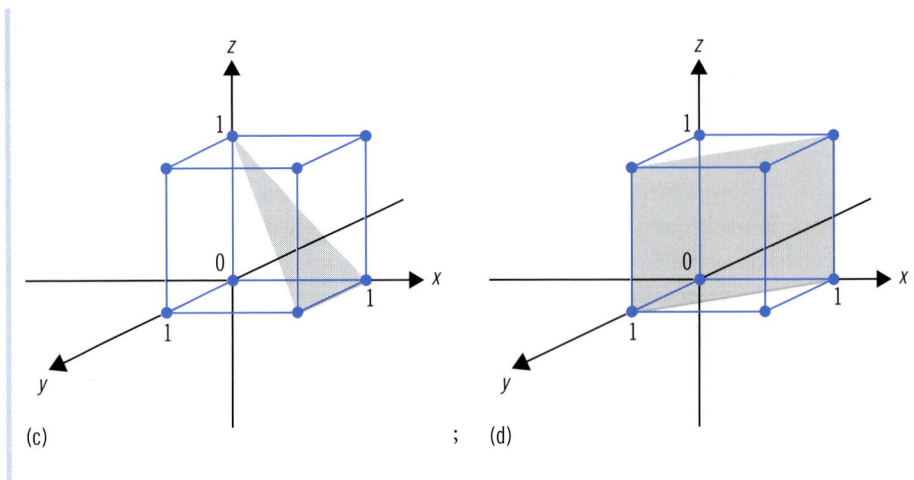

(c)   ;   (d)

2.3 X-RAY DIFFRACTION

Section learning outcomes

To be able to:

- Describe and summarize how X-rays are generated;
- Identify X-rays suitable for an X-ray diffraction experiment;
- Discuss how X-rays are diffracted by a crystal lattice;
- Illustrate and explain Bragg's law.

2.3.1 X-rays

Röntgen named his discovery 'X-ray' because *x* represented an unknown quantity in mathematics.

X-rays were first discovered in 1895 by Wilhelm Conrad Röntgen, [1] who was awarded the 1901 Nobel Prize in Physics for his discovery. Essentially, X-rays are a type of radiation that have short wavelengths and are very high in energy.

The wavelengths of X-rays are typically between 0.1 and 100 Ångstroms ($\text{Å}; 1\ \text{Å} = 10^{-10}\text{m}$). This is of the same order as typical molecular bond distances, which lie in the range of 0.8 to 3.0 Å, and this enables X-rays to interact with the contents of a crystal, allowing a crystallographic experiment to be performed.

X-rays are generated when an electron in an excited state relaxes back to its ground state. The excited state can usually be induced by changing the velocity of a fast-moving electron. Figure 2.3 shows that when a photon collides with an electron in an atom, the electron is promoted into an excited state.

Energy is emitted when this excited state relaxes back to its ground state; this energy includes X-ray irradiation. The total amount of energy emitted is the equivalent of the energy difference between the excited state and the ground state; and also corresponds to the specific wavelength of the emitted radiation, where $E = h\nu$ (E is the energy emitted, h is Planck's constant and ν is the frequency of the wavelength).

When an electron relaxes to its ground state from an excited state, the types of radiation that are emitted depend on which atomic shells are involved in the transition. The atomic shells, K, L, and M are represented in Fig 2.4. The figure also shows the types of radiation that are emitted when the electron moves between these shells. When an electron moves from L to K, it gives $K\alpha$ (both type 1 and 2) radiation; when it goes from M to K it gives $K\beta$ radiation and when it moves from M to L it gives $L\alpha$ radiation. The wavelengths of $K\alpha$ are longer than $K\beta$.

For an X-ray diffraction experiment, we are generally interested with the $K\alpha$ wavelength. The $K\alpha$ wavelength is known as the 'characteristic X-ray' of a given substance and it changes from element to element since each has different energy levels.

TABLE 2.2 Radiation wavelength in Ångstroms (Å)

	Copper	**Molybdenum**
Kα_1	1.54433	0.71354
Kα_2	1.54051	0.70962
Kβ	1.39217	0.63225

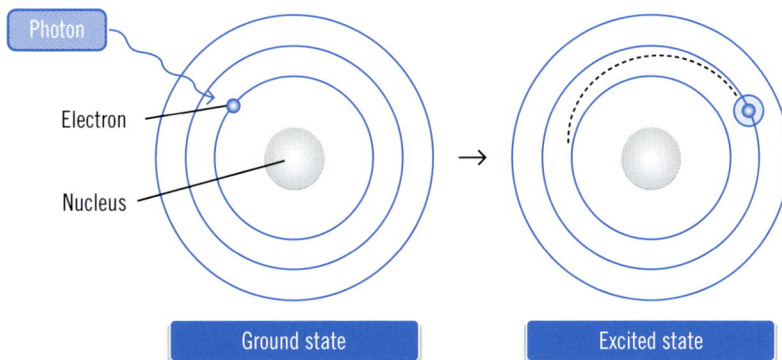

FIGURE 2.3 Emission of X-rays

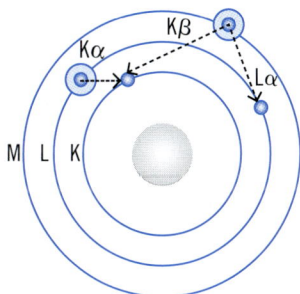

FIGURE 2.4 Radiation types

Copper and molybdenum are the two most common sources used for X-ray diffraction experiments. Table 2.2 summarizes their radiation wavelengths in Ångstroms (Å; 1 Å = 10^{-10} m)

The X-rays used for a diffraction experiment consist of the combination of $K\alpha_1$ and $K\alpha_2$, which are not usually resolved, while $K\beta$ radiation is usually eliminated using monochromators. $K\alpha$ radiation wavelengths for X-rays are typically **0.71073 Å for Mo** and **1.5418 Å for Cu**.

As atom types each have characteristic absorption wavelengths, a particular element can be used to filter out the $K\beta$ radiation by absorbing it. For example, Zr is used to filter out $K\beta$ radiation for Mo, while Ni does the same for Cu. However, this will also absorb some of the $K\alpha$ and decrease its intensity.

Refer to Chapter 8 for more details on the X-ray diffraction experiment.

2.3.2 Diffraction by crystals

During an X-ray diffraction experiment, a crystal is irradiated with X-rays. The interaction between the oscillating electrons within each atom and the incoming X-ray beam causes the X-rays to diffract in all directions. The diffracted X-ray beams from all the Miller sets gives rise to '**diffraction spots**'. The diffraction spots represent the **diffraction pattern** of the crystal being examined.

The sketch in Fig. 2.5 outlines the process of an X-ray diffraction experiment. When the incident X-rays from the X-ray source reach the crystal, the electrons in each Miller set diffract the X-rays accordingly, giving rise to diffraction spots. The diffraction of X-rays by a crystal lattice can be explained by Bragg's law.

2.3.3 Bragg's law

Bragg's law is able to provide a simplified explanation as to how diffraction spots and patterns are produced from the incident X-ray beam.

Looking at Figure 2.6, in a crystal lattice, the lattice lines and planes (Miller sets) are located at a separation distance of d (this is also known as d-spacing). When the incident X-rays collide with the crystal lattice at specific points, the X-rays are reflected or diffracted out from these points giving rise to diffraction spots.

The path taken by the incident and reflected beam can be calculated as the sum of AY and YB and this is equal to the wavelength of the incident X-ray beam, $n\lambda$. If the angle at which

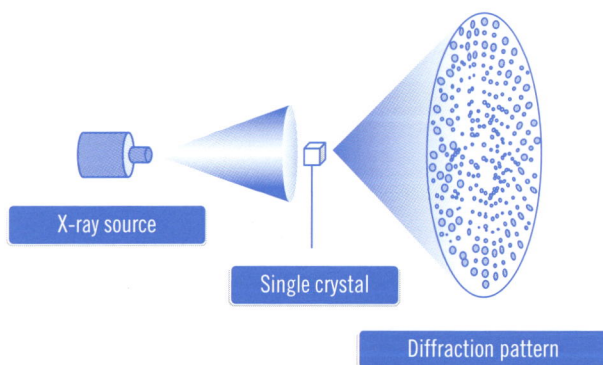

X-ray source

Single crystal

Diffraction pattern

FIGURE 2.5 Cartoon representation of a single crystal X-ray diffraction experiment

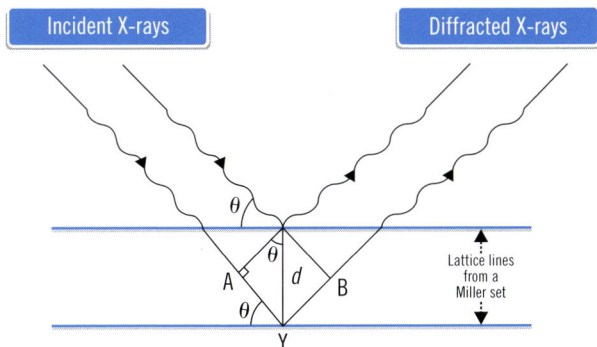

FIGURE 2.6 Diffraction according to Bragg's law

the incident beam collides with the crystal is θ, then we can calculate the path difference in terms of d, θ, and λ.

$$AY + YB = n\lambda,$$

and we know that

$$\sin\theta = AY/d,$$

rearranging the equation

$$AY = d\sin\theta \quad (AY = YB),$$

therefore, the path difference is

$$2d\sin\theta = n\lambda.$$

In summary, Bragg's law states that diffraction spots occur when $2d\sin\theta = n\lambda$, that is, the incident X-ray wavelength is twice the d spacing multiplied by the sine of the incident beam angle.

SELF-TEST QUESTIONS

1. Write brief notes on the following:

    (a) Why are X-rays suitable for single crystal diffraction experiments?

    (b) What are the most common sources of X-rays?

    (c) Explain how X-rays are produced.

    (d) Outline the radiation types produced from electron movement between atomic shells.

2. Describe briefly, with the aid of diagrams, how a diffraction pattern can be produced.

3. (a) With the aid of diagrams, explain how a crystal diffracts X-rays according to Bragg's law.

    (b) Also, prove that $2d\sin\theta = n\lambda$.

2.4 DIFFRACTION PATTERNS AND RECIPROCAL SPACE

Section learning outcomes

To be able to:

- Understand how diffraction patterns are produced;
- Understand how Miller indices relate to diffraction spots;
- Relate the concepts of real space and reciprocal space.

2.4.1 Diffraction patterns

In the earlier sections, we have examined how X-rays interact with the crystal lattice and how this interaction can be explained by Bragg's law. In this section, we look at the consequence and results of that interaction.

If we refer again to Fig. 2.5, we see that in an X-ray diffraction experiment, the incident X-ray beam is diffracted by the crystal lattice. The diffracted X-rays are stopped by a data collection plate (previously photographic film was used, modern diffractometers today usually use CCDs or image plates; this is detailed further in Chapter 8). The diffracted beams are observed on the collection plate as spots – also known as diffraction spots.

Each diffraction spot is the sum result of the diffraction from a Miller plane, thus each spot can be identified with an *hkl* value corresponding to the Miller indices. Each spot also has an intensity value, depending on how strong or weak the resulting diffracted X-ray is. In some cases, a spot may be *absent* when the intensity corresponds to zero. These are known as *absences*. (This is explained further in Chapter 5.)

Each set of diffraction spots produced and collected is known as a *frame* of data. In the course of a typical experiment, several thousand frames may be collected. These frames of data can then be reconstructed to produce a *diffraction map*, and subsequently an electron-density map, of the crystal lattice. This diffraction map in essence consists of the reciprocal form of the crystal lattice under study.

2.4.2 The reciprocal lattice

The processing and 'reconstruction' of all the frames of diffraction data produces the reciprocal form of the crystal under study.

To understand this concept better, if we consider Bragg's law (Section 2.3.3), where

$$2d\sin\theta = n\lambda,$$

the equation can be rearranged as

$$\sin\theta = \left(\frac{n\lambda}{2}\right)\left(\frac{1}{d}\right).$$

The equation shows that $\sin\theta$ is inversely related to d; from this we can infer that when a lattice has large d values the diffraction pattern will be compressed, while a lattice with small *d* values will have a diffraction pattern that is well spread out. Experimentally, when a lattice has long axes the diffraction spots tend to appear very close together while if a lattice has short axes, then the spots appear to be more spread out.

We can also say that $\sin\theta$ is directly related to the reciprocal of *d*, that is, $1/d$. If we recall that *d* represents the spacing between the sets of Miller planes, then we see that the resultant diffraction pattern is in units of $1/d$, this is known as reciprocal space. If we assume that for each set of Miller planes (each value of *hkl*) in real space there is a point in reciprocal space, then the collection of all of these points can be referred to as the reciprocal lattice. This relationship is illustrated in Fig. 2.7.

A mathematical correlation between real and reciprocal space can be given by the following equation. Assuming that the starred values are that of the reciprocal then:

$$a.a^* = b.b^* = c.c^* = 1; \alpha = \alpha^*; \beta = \beta^*; \gamma = \gamma^*.$$

In the X-ray diffraction of crystals, we experimentally measure the intensities of points in reciprocal space during data collection. These data are ultimately used to produce an electron-density map for a portion of the unit cell in real space for a given crystal, from which the structure can be solved and refined.

Table 2.3 outlines the correlations between real space and reciprocal space and all its other equivalents.

The term *real space* used here to denote the physical crystal lattice is also sometimes known as *direct space*.

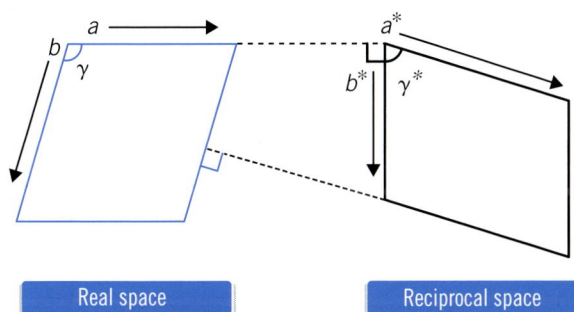

FIGURE 2.7 A sketch outlining the relationship between real space and reciprocal space

TABLE 2.3 Correlations between real space and reciprocal space

Real space	Reciprocal space
Crystal structure	Diffraction pattern
Electron density (atomic parameters)	Amplitudes and phases (X-rays)
Crystal lattice; unit cell	Reciprocal lattice; cell
Coordinates (*x*, *y*, *z*)	Coordinates (*h*, *k*, *l*)

SELF-TEST QUESTIONS

1. With the aid of Bragg's law, explain why a crystal lattice with short axes would have a diffraction pattern in which spots would appear far apart.

2. Describe, with the use of diagrams, how real space is related to reciprocal space.

◘ CHAPTER SUMMARY

1. Weiss indices provide an alternative method of labelling or notating cell parameters, particularly against a Cartesian plot.

2. A crystal lattice can be further divided into imaginary planes known as Miller planes. These can be identified by integers represented by values of h, k, and l. These (h, k, l) values are known as Miller indices.

3. The equivalence of Cartesian coordinates, Miller indices, and unit-cell dimensions is:

Cartesian coordinates	x	y	z
Unit-cell dimensions	a	b	c
Miller indices	h	k	l

4. Miller indices can be used to provide a quick visual representation of the planes in a crystal lattice:

   (a) The smaller a Miller index is, the more parallel the plane is to the axis;

   (b) The larger a Miller index is, the more perpendicular the plane is to that axis;

   (c) If a Miller index is zero, the plane is parallel to that axis.

5. Miller indices are determined from the intercepts of the planes along the axes, taking the reciprocals, clearing the fractions, and finally reducing the indices to the lowest terms.

6. X-rays are able to interact with the contents in a crystal because the wavelength of the X-rays (0.1–$100\,\text{Å}$) are of the same order as the typical molecular bond distances (0.8–$3.0\,\text{Å}$)

7. Copper ($\lambda = 1.54\,\text{Å}$) and molybdenum ($\lambda = 0.71\,\text{Å}$) are the two most common sources of X-rays used in diffraction experiments.

8. When a crystal is irradiated with X-rays, the atoms in each set of Miller planes (Miller sets) in the crystal lattice diffract the X-rays, producing diffraction spots.

9. In an X-ray diffraction experiment, the crystal diffracts X-rays according to Bragg's law.

10. Bragg's law states that $2d\sin\theta = n\lambda$.

11. When all the frames of diffraction data are processed and reconstructed, the reciprocal form of the crystal under study is obtained.

12. The crystal lattice is related to the reciprocal lattice by the equation $a.a^* = b.b^* = c.c^* = 1$; $\alpha = \alpha^*$; $\beta = \beta^*$; $\gamma = \gamma^*$.

REFERENCES

1. The Nobel Foundation. (1967). *Nobel Lectures, Physics 1901–1921.* Elsevier, Amsterdam.

FURTHER READING

Giacovazzo, C., Monaco, H. L., Artioli, G., *et al.* (eds.) (2002). *Fundamentals of Crystallography,* 2nd edn. IUCr Texts on Crystallography, Oxford University Press, New York.

Hammond, C. (2001). *The Basics of Crystallography and Diffraction.* IUCr Texts on Crystallography, Oxford University Press, New York.

LINKS

Online learning tools:

Generating your own diffraction pattern: http://math.arizona.edu/~ura/964/sinclair/diff.html

A Moseley plot of the characteristic X-rays: http://hyperphysics.phy-astr.gsu.edu/hbase/quantum/moseley.html

Draw your own lattice planes: http://www.doitpoms.ac.uk/tlplib/miller_indices/lattice_draw.php

SYMMETRY ELEMENTS

3

By the end of this chapter you should be able to:

- Gain an appreciation of the forms of symmetry that can occur in a crystal lattice;
- Recognize the translational and non-translational symmetry elements;
- Be familiar with the Bravais lattice types and understand how various symmetry elements may occur in each;
- Understand that symmetry elements can influence diffraction data.

3.1 INTRODUCTION

We learnt in Chapter 1 how a unit cell may be further reduced into smaller components, known as asymmetric units. Within a unit cell, each and every asymmetric unit is related by symmetry. In this chapter we examine the possible types of symmetry that are associated with each asymmetric unit.

3.2 LATTICE POINT SYMMETRY

Section learning outcomes

To be able to:

- Identify how asymmetric objects are related by symmetry elements within a unit cell;
- Categorize translational and non-translational symmetry elements;
- Recognize the non-translational symmetry elements that cause absences in diffraction data;
- Visualize and model the symmetry operations that can occur in crystallography.

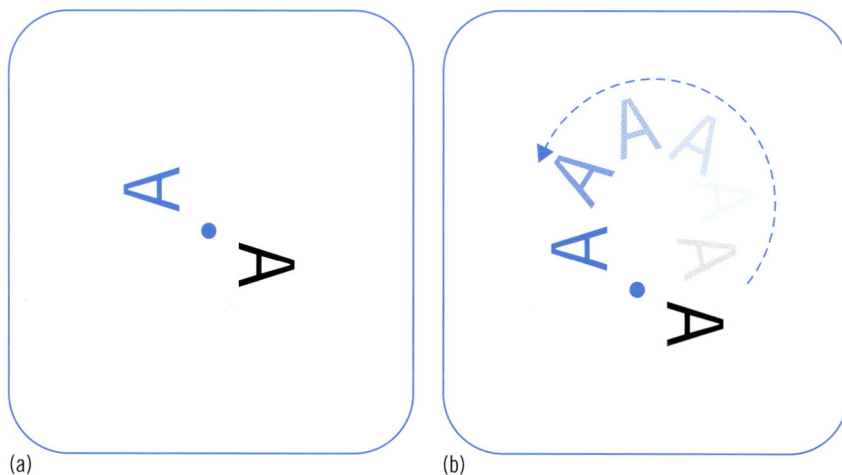

(a)   (b)

FIGURE 3.1 (a) Two objects related by a symmetry operation; (b) the object undergoing a 'movement' from one position to another

3.2.1 Asymmetric objects within a unit cell

Typically our comprehension of symmetry is such that we assume two asymmetric units to be related by a form of symmetry, however it often aids understanding and learning if we can imagine an object (usually the object within the asymmetric unit) undergoing a 'movement' from its original position to a new position following a 'pathway' (in this case, a symmetry operation), as in Fig. 3.1.

It is important to remember at this point that while crystallography examines molecules in three dimensions, we are often limited to conveying crystallographic data and material in two-dimensional form, that is, on paper. Therefore, to aid this understanding, general notations and symbols are used to represent the symmetry elements (these are examined further in Chapter 4).

Some symbols depend on whether the view is taken on a plane view (plan view, from top), where the symmetry axes are normal to the plane (mainly given in the following examples) or a perpendicular view (side-on view), where the symmetry axes are parallel to the plane of projection. These are explained further in the following sections.

Note: For clarity, we shall distinguish between a *symmetry element* and a *symmetry operation*. A symmetry operation is the process of moving an object to an equivalent position while a symmetry element is the geometrical entity (line, point, or plane) about which the symmetry operation occurs.

3.2.2 Types of symmetry element

The types of symmetry element that occur in crystal lattices can broadly be divided into two large groups: non-translational and translational symmetry elements. In diffraction data, these can be differentiated by the occurrence of absences. Absences in diffraction data occur when the intensity corresponding to a specific Miller index (*hkl* value) is zero.

Translational symmetry elements can be identified by a translation. A translation of an object can be defined as moving an object in a direction (a, b, c) without rotating or reflecting the object. For example an object located at (x, y, z) when translated by (a, b, c) will be located at $(x + a, y + b, z + c)$. These types of symmetry elements **will** cause absences within diffraction data. These are screw axes and glide planes.

Absences or absent reflections in diffraction data occur when the observed intensity $I^{obs} = 0$.

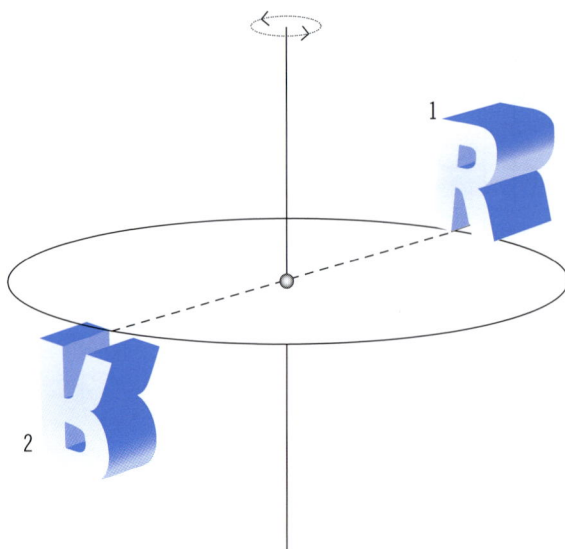

FIGURE 3.2 The inversion centre

Non-translational symmetry further incorporates a form of rotation of the object. Non-translational symmetry elements **do not** cause absences in crystal data. These are centre of symmetry (inversion centre), reflection, rotation (only two-, three-, four-, or six-fold), and rotary inversion. The following sections describe these symmetry elements in detail.

Non-translational symmetry elements

The four non-translational symmetry elements do not cause absences in diffraction data.

The inversion centre

An inversion centre or a centre of symmetry is typically identified by a bold or darkened point. It is also denoted in writing with a bar across the top of a number, e.g., $\bar{1}$.

Referring to Fig. 3.2, an object located at 1 is inverted through the centre of inversion to a new location at 2. This resulting object is the mirror image of the original, turned upside-down.

Learning tip: **If we were to imagine the inversion centre to be a deep dark pit, then when the object at 1, falls head-first downwards through that pit, it emerges on the direct opposite side, still on its head, thereby relocating to 2, but inverted to its mirror image.**

Reflection

A reflection, like the most common reflections that we know of, takes place across a mirror plane (see Fig. 3.3). The mirror plane is denoted by the letter **m** and by a dark horizontal line in the perpendicular view (—) while in the plane view it is denoted by a top-right corner (⌐).

An object located at 1 undergoing a reflection would be reflected to the opposite side at 2, at equidistance from the mirror plane.

Learning tip: **Your right hand is the mirror image of the left and if you held up both your hands, facing each other, for example in preparation to clap, midpoint between your two hands would be the mirror plane.**

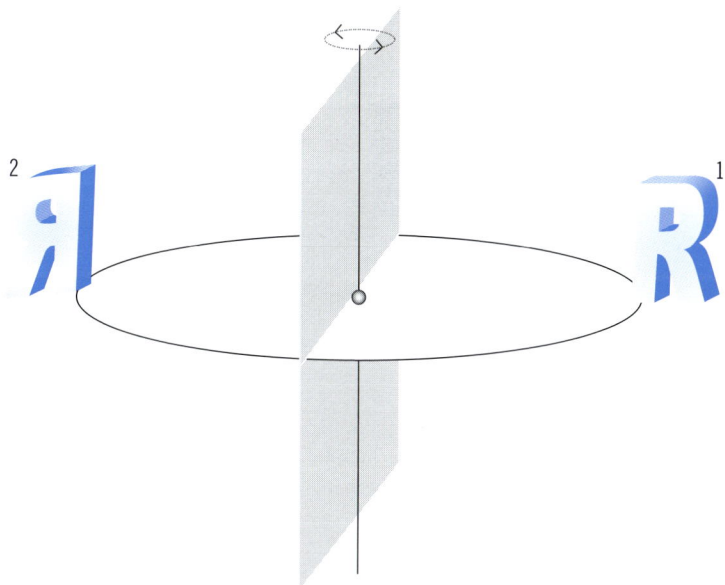

FIGURE 3.3 The mirror plane

TABLE 3.1 Typical rotations and symbols where axes are normal to plane

Name of rotation	Notation	Symbol
Diad	2-fold	◆
Triad	3-fold	▲
Tetrad	4-fold	◆
Hexad	6-fold	●

Rotation (only one-, two-, three-, four-, or six-fold)

All rotations in crystallography occur *counterclockwise*, as a fraction of the circle upon which that rotation occurs. There are only five types of rotation; one-, two-, three-, four-, and six-fold. Rotations are denoted by an integer, n (n-fold) and by the symbols given in Table 3.1.

Figure 3.4 shows an example of a four-fold rotation. The angle of rotation is determined by the fraction, $360/n$; this means that each object is rotated by $360/4 = 90°$.

For example, an object at 1 is rotated counterclockwise by 90° to its new position at 2. A further rotation of another 90° moves the object from 2 to 3 and subsequently to 4. Visually, the object at 3 is the mirror image of the object at 1, while the object at 4 is the mirror image of that at 2.

Rotation-inversion

A rotation-inversion consists of a combination of two symmetry elements; a rotation followed by an inversion. Here again the rotations occur in a *counterclockwise* fashion, and are considered as a fraction of a full circle. A rotation-inversion is denoted by a bar across the top of the allowed rotations, $\bar{2}$, $\bar{3}$, $\bar{4}$, or $\bar{6}$, and by the symbols in Table 3.2.

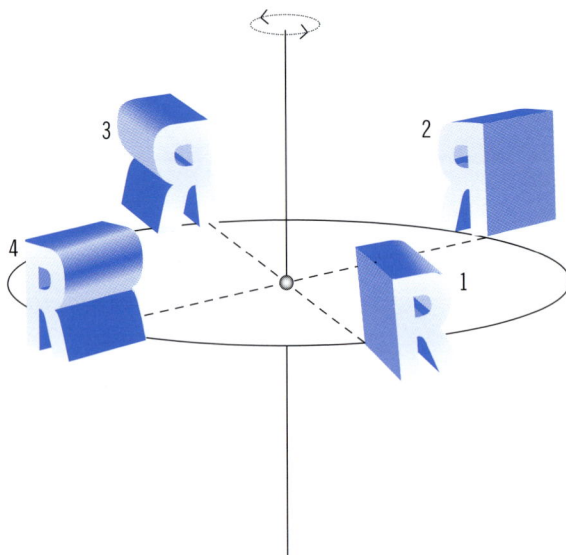

FIGURE 3.4 A Four-fold rotation

TABLE 3.2 Typical rotation-inversions and symbols, where axes are normal to plane

Name of rotation-inversion	Notation	Symbol
Diad	2-fold	◊
Triad	3-fold	▲
Tetrad	4-fold	◆
Hexad	6-fold	●

The example in Fig. 3.5 shows a four-fold rotation-inversion, denoted by $\bar{4}$ and by the symbol ◆. Here again, each object is rotated *counterclockwise* by 360/n, 90°. Referring to Fig. 3.5, the object at 1 is rotated by 90° to the position at 4 and then inverted through a centre of inversion in the midpoint to 2. Following this, the object at 2 is again rotated by 90° to its position at 1 and then inverted through the inversion centre to position 3 (the right way up).

From its position at 3, the object is again rotated counterclockwise by 90° to 2 (still the right way up) and then inverted through the centre to its final position at 4.

Translational symmetry elements

There are only two types of translational symmetry in crystallography; the screw axis and the glide plane. Both of these can be identified in diffraction data by absences caused.

Screw axis

A screw axis consists of a combination of two symmetry operations; a translation followed by a rotation. Again, the rotations occur *counterclockwise*. A screw axis is denoted by two

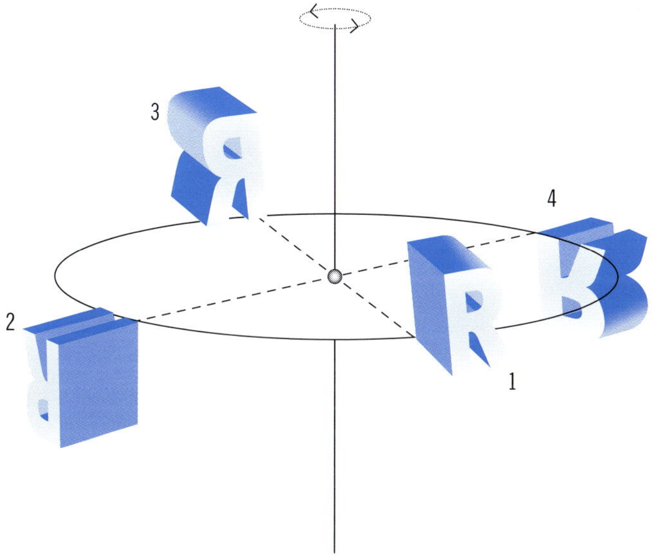

FIGURE 3.5 A four-fold rotation-inversion

TABLE 3.3 Types of screw axis and representative symbols

Name of screw axis	Notation	Symbol
Diad	2_1	⸮
Triad	$3_2, 3_1$	▲ ▲
Tetrad	$4_1, 4_2, 4_3$	✦ ✦ ✦
Hexad	$6_1, 6_2, 6_3, 6_4, 6_5$	✦ ✦ ✦ ✦ ✦

integers, the first representing the value of the rotation (limited to two, three, four, and six), while the second is written in subscript form (a value smaller than the first) and in part represents the fraction of translation. This can be written generally as M_n, where each object is rotated by $360/M$ degrees, then translated upwards by a fraction n/M of the unit cell.

Table 3.3 lists all the possible types of screw axis and their symbols.

The example shown in Fig. 3.6 represents two possible types of three-fold screw axis; (a) 3_1 and (b) 3_2.

Examining first the example of the 3_1 screw axis, each object is rotated through $360/3 = 120°$ and then translated upwards by 1/3 of the unit cell axis. The object located at 1 is first rotated through 120° on the same plane and then translated upwards by 1/3 of the axis to a new location at 2. From 2 the object is again rotated 120° and then translated upwards by another 1/3 of the axis, moving it to 3. From 3, the object again rotates 120° and is translated upwards by 1/3, returning it to its original position although translated upwards by a full length of the unit cell.

In the example of the 3_2 screw axis, each object is similarly rotated through 120°; however each is translated upwards by 2/3 of the unit cell axis. Starting in the same position at 1, the object is rotated through 120° and then translated upwards 2/3 of the axis to its new

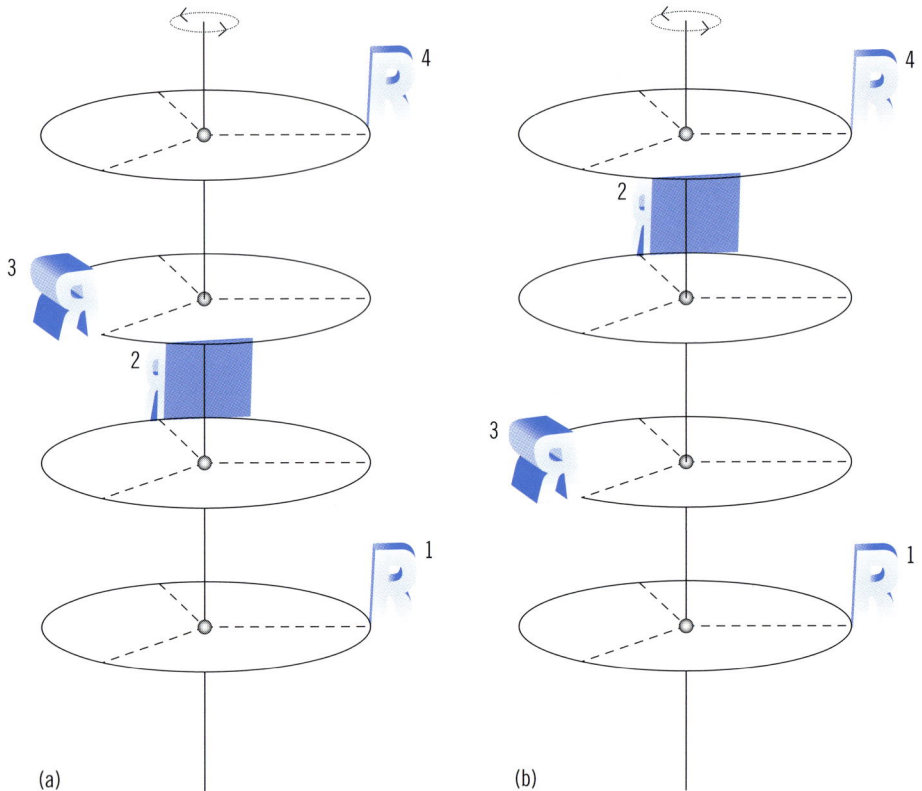

FIGURE 3.6 Screw axes: (a) 3_1 screw axis; (b) 3_2 screw axis

position at 2. From 2, it is rotated another 120° and then translated upwards another 2/3 of the axis, effectively bringing it to its new position at 3, in the unit cell above. The object is similarly returned to its original position translated by the length of one cell axis when the object at 3 is rotated through 120° and then translated another 2/3 upwards to 4.

Screw axes are so named because if we were to look down the midpoint of each diagram and follow the rotation of the objects visually, the 3_1 screw axis would resemble an anticlockwise corkscrew while in the 3_2 screw axis, the objects would resemble a clockwise corkscrew.

Glide planes

A glide plane also consists of a combination of two symmetry operations, in this case a translation followed by a reflection. Glide planes are denoted by the planes along which the glide occurs, most commonly a, b, c, or n (an n glide typically refers to a diagonal direction within the unit cell). In diagrams they are represented by various dotted lines depending on whether they are parallel (-------), perpendicular to the plane (·······) or diagonal to the plane (– · – · – · –). When an object undergoes a 'glide', it is translated by half the length of the cell axis, and then reflected to its mirror image.

In the example shown in Fig. 3.7, the object at 1 is translated by half the axis length and reflected to its mirror image at 2. From 2 it is again translated by a half of the axis length and then reflected again to 3, where it is essentially the original object translated by the full length of the axis.

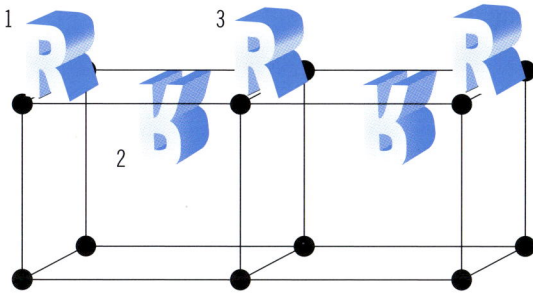

FIGURE 3.7 Objects undergoing a glide

SELF-TEST QUESTIONS

1. How can translation and non-translational symmetry elements be differentiated?

2. Name the types of symmetry elements to be found in each group.

3. With the aid of diagrams, describe a 2_1 screw axis.

Group learning activity – modelling the symmetry operations

To aid the visualization and understanding of the symmetry operations, it is useful to be able to model and subsequently discuss how the symmetry elements work. The following are instructions for folding a three-dimensional modular origami object. Working in groups of four or five, using two different coloured sheets of paper, between 10–15 cm (4–6 in) squares, each person creates one of these modular objects.

   To create an asymmetric object, fold the bottom edge of the completed object to the right. Ensure that all the objects in the group are the same. Referring to Section 3.2, model the

Making the module

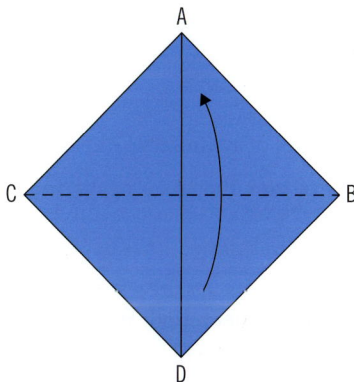

1. Crease and fold a vertical diagonal then fold D up to A.

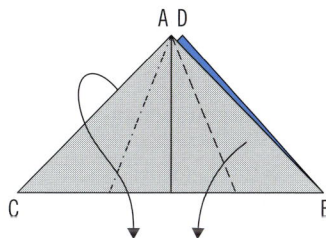

2. Fold edge AD, B forward to the vertical crease (valley fold) and edge AD, C behind to that crease (mountain fold).

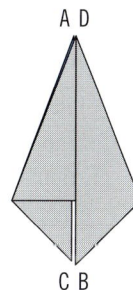

3. This is the completed module. Make another.

Assembling the module

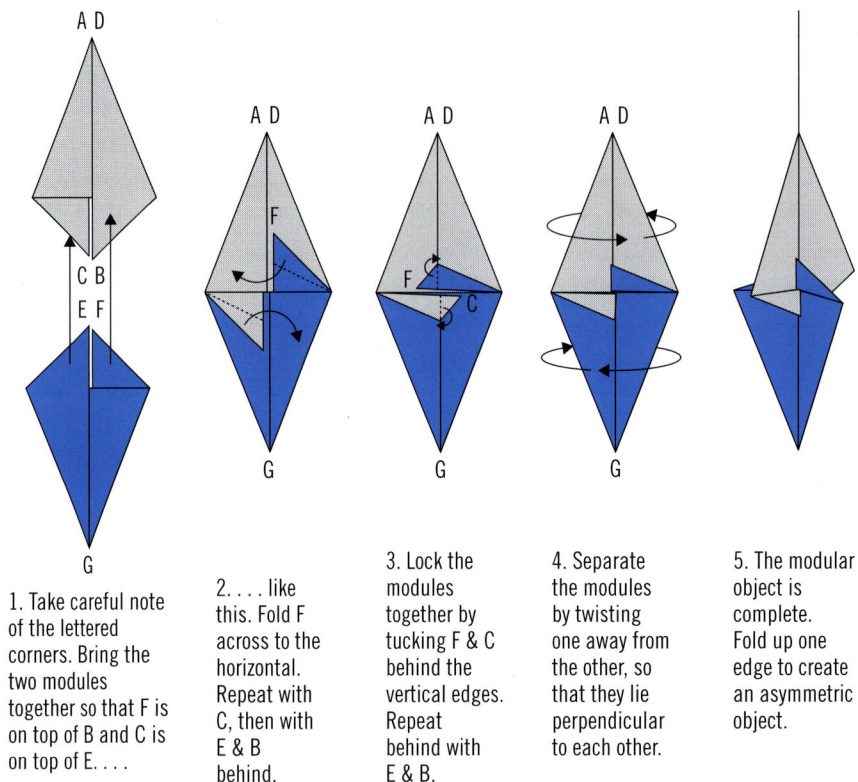

1. Take careful note of the lettered corners. Bring the two modules together so that F is on top of B and C is on top of E. . . .

2. . . . like this. Fold F across to the horizontal. Repeat with C, then with E & B behind.

3. Lock the modules together by tucking F & C behind the vertical edges. Repeat behind with E & B.

4. Separate the modules by twisting one away from the other, so that they lie perpendicular to each other.

5. The modular object is complete. Fold up one edge to create an asymmetric object.

movement and end results of the symmetry operations. This origami object was designed by Paul Jackson.[1]

3.3 LATTICE TYPES AND SYMMETRY ELEMENTS

Section learning outcomes

To be able to:

- Recognize the association between lattice types and types of symmetry;
- Extrapolate the possible number of symmetry operations that may occur in a particular lattice type.

We have so far identified the six possible types of symmetry element, both translational and non-translational, that can occur in crystals. In this section, we will see how we can determine the types of symmetry operations that may occur in the different lattice types.

---

[1] Module instructions used by permission of Paul Jackson, designer and copyright holder.

Not all of the types of symmetry operator can arise in all of the 14 Bravais lattice types. The types of symmetry operation that can be found in each lattice type are restricted by the shape of the 'lattice box' defined by its axes and angles. For example, it would be impossible for a mirror plane to occur in a triclinic lattice that has no equivalent axes and no faces that lie at right angles.

However, in the higher symmetry lattices it becomes possible and more likely for one or more symmetry operations to occur in the lattice space. It is the combination of the number of equivalent axes and number of right angles within a lattice type that influences the symmetry operations that can occur. In each case, a symmetry operator occurs along a specific axis of the lattice. (Note: symmetry operations can also sometimes occur along a diagonal within the unit cell – these are described as taking place along the n-direction.)

In the following sections, we shall examine the lattice types and consider the symmetry operations that can be found in each.

3.3.1 Triclinic lattices ($a \neq b \neq c$; $\alpha \neq \beta \neq \gamma \neq 90°$)

As shown in the figure, in a triclinic lattice all of the axes are of different lengths and none of the faces is at right angles to any other. When we consider this, it means that there is a distinct lack of symmetry within the lattice space and therefore the choice of symmetry elements that may be found in the triclinic lattice become limited. The only form of symmetry that may occur is the centre of symmetry or the inversion centre.

3.3.2 Monoclinic lattices ($a \neq b \neq c$; $\alpha = \gamma = 90°$, $\beta \neq 90°$)

In a monoclinic lattice, while there are two right angles within the lattice, all of the axes are of different lengths. Within this type of lattice there is usually a minimum of a two-fold rotation or rotation-inversion.

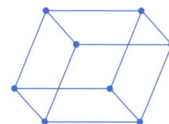

3.3.3 Trigonal lattices ($a = b \neq c$; $\alpha = \beta = 90°$, $\gamma = 120°$)

A trigonal lattice generally consists of a diamonoid shape, where the height of the lattice is of a different length from the other two axes, which are equivalent. Two of the lattice angles are at 90° while the third is at 120°. In the trigonal lattice, the minimum symmetry element that can occur is one three-fold rotation or rotation-inversion.

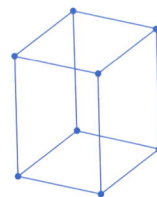

3.3.4 Hexagonal lattices ($a = b \neq c$; $\alpha = \beta = 90°$, $\gamma = 120°$)

Visually, hexagonal lattices are an elongated hexagonal shape with the equivalent dimensions where the axes on the hexagon are equal lengths and the height is different. The angles on the hexagonal face are at 120°, while the others are at right angles. In a hexagonal lattice type, the minimum symmetry element that can occur is one six-fold rotation or rotation-inversion.

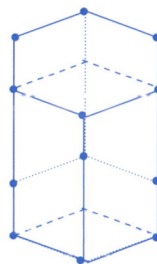

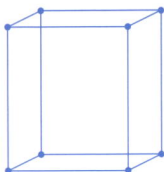

3.3.5 Orthorhombic lattices ($a \neq b \neq c$; all angles $= 90°$)

The orthorhombic lattice has a box shape in that all the faces lie at right angles, while the axes' lengths differ. In the orthorhombic lattice, the minimum symmetry operations that can occur are three two-fold rotations or rotation-inversions.

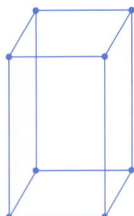

3.3.6 Tetragonal lattices ($a = b \neq c$; all angles $= 90°$)

The tetragonal lattice also has a box shape in that all the faces lie at right angles. Two of the axes are of equal length. In the tetragonal lattice, the minimum symmetry operation that can occur is a four-fold rotation or rotation-inversion.

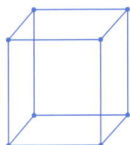

3.3.7 Cubic lattices ($a = b = c$; all angles $= 90°$)

The cubic lattice has a box shape with all the faces at right angles and all axis lengths equal. In the cubic lattice, the minimum symmetry operations that can occur are four three-fold rotations or rotation-inversion.

SELF-TEST QUESTIONS

1. Explain why not all types of symmetry operations can occur in all of the Bravais lattice types.

2. Owing to the dimensions of a triclinic lattice, only limited types of symmetry element can occur; briefly explain why and which ones.

3. Explain why a screw axis cannot occur in a triclinic lattice.

4. How many symmetry operations might you expect to find in an orthorhombic lattice?

◉ CHAPTER SUMMARY

1. Within each unit cell, the asymmetric units are related by symmetry elements.

2. The symmetry elements can be divided into two categories: translational (screw axes, glide planes) and non-translational (inversion centre, mirror plane, rotation, rotation-inversion).

3. Translational symmetry elements cause absences in diffraction data.

4. The type and number of symmetry elements that can occur in crystal lattices are limited by the lattice dimensions.

📖 FURTHER READING

Giacovazzo, C., Monaco, H. L., Artioli, G., *et al.* (eds.) (2002). *Fundamentals of Crystallography,* 2nd edn. IUCr Texts on Crystallography. Oxford University Press, New York.

Glusker, J. P. and Trueblood, K. N. (1985). *Crystal Structure Analysis – A Primer.* Oxford University Press, New York.

Hahn, T. (ed.) (2002). *International Tables for Crystallography: Brief Teaching Edition of Volume A (Space-group Symmetry),* 5th edn. Kluwer Academic, Dordrecht.

LINKS

Online learning tools
Experiment with space groups and symmetry: http://escher.epfl.ch/cse

SPACE GROUPS

4

By the end of this chapter you should be able to:

- Recognize and understand the Hermann–Mauguin notation;
- Understand and appreciate the use of space-group diagrams;
- Identify the symmetry elements represented in the Hermann–Mauguin notation;
- Identify atoms in special positions;
- Distinguish between centrosymmetric and non-centrosymmetric space groups.

4.1 INTRODUCTION

After examining and understanding the symmetry elements and how they can be found in connection to the various Bravais lattice types in Chapter 3, we now need to consider how this information may be easily shared or transmitted. Crystallographers use a simple and concise mathematical representation of lattice types and symmetry operators, known as space groups. Here, we will look at the concept of space groups and examine why there are 230 space groups. We will learn the use of the Hermann–Mauguin notation and space group symbols in representing symmetry operations.

4.2 SPACE GROUPS

Section learning outcomes

To be able to:
- Understand and interpret the Hermann–Mauguin symbols;
- Understand the association of Hermann–Mauguin symbols along the axes;
- Read and interpret simple space-group diagrams.

Earlier, at the end of Chapter 3, we examined how not all symmetry operators could occur in all of the Bravais lattice types. We found that the higher the symmetry of the Bravais lattice, the more likely it was to contain one or more symmetry elements. The type of centring of the Bravais lattice (primitive, body-centred, or face-centred) also contributes significantly to the symmetry elements that may be found within that lattice type. Various combinations of the six symmetry operations with the different lattice types produce 230 unique arrangements of objects that fill space; hence the 230 space groups distributed across seven crystal systems.

All of the 230 space groups are collected in a book known as the *International Tables for Crystallography.* They are represented by the symmetry-elements diagram and general position diagram, notated with the Hermann–Mauguin symbols and other relevant information. These are explained further in the following sections. *International Tables* serves both to list all occurring space groups and to act as a reference in identifying the space groups. Table 4.1 shows a breakdown of the number of space groups that can be found in each crystal system.

TABLE 4.1 Number of space groups in each crystal system

Crystal systems	Number of space groups
Triclinic	2
Monoclinic	13
Orthorhombic	59
Tetragonal	68
Trigonal	25
Hexagonal	27
Cubic	36
Total	**230**

4.2.1 Hermann–Mauguin notation

If we imagine that the Bravais lattice types and symmetry notation form the alphabet of the crystallography language then the Hermann–Mauguin notation combines the two to form the coherent words.

The Hermann–Mauguin notation is written with the first letter representing the lattice type and the following three notations representing the type of symmetry operation or operations along the corresponding axes of x, y, and z.

For example, the orthorhombic space group P is notated as $2_1 2_1 2_1$,

Direction	x	y	z
Lattice	a	b	c
Space group	$P 2_1 2_1 2_1$		

which explains that the lattice is a primitive lattice with a 2_1 screw axis along each of the x, y, and z directions.

In the case of glide planes, however, the notation is such that the named glide plane runs *perpendicularly to* the axis.

For example in the monoclinic space group, $P\,1c\,1$, which is usually shortened to $P\,c$,

Direction	x	y	z
Lattice	a	b	c
Space group	$P\,1c\,1$		

explains that the lattice is a primitive lattice with a c glide plane perpendicular to the b-direction, while there are no other symmetry elements along the other directions.

Visually the c glide perpendicular to b is as follows. In Fig. 4.1, the dotted plane shows the glide plane that lies perpendicular to the y direction. Beginning with the original object located at (x, y, z) the object undergoes a translation along half the unit cell in the direction of z, resulting in object 1 $(x, y, z + \frac{1}{2})$. The object moves to its final position, undergoing a reflection across the glide plane resulting in object 2 $(x, -y, z + \frac{1}{2})$.

The Hermann–Mauguin notations also correspond to two-dimensional diagrams used to pictorially represent the symmetry elements and objects within the space groups. These are the space-group diagrams.

4.2.2 Space-group diagrams

When we learn to think in terms of three dimensions, we are able to project the third dimension visually in our mind. However, as one person's visual image is unlikely to be exactly the

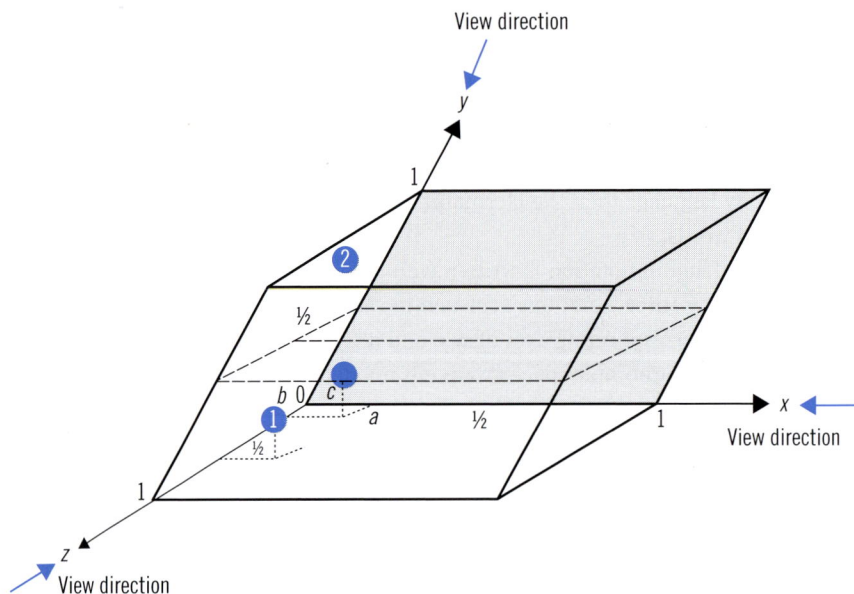

FIGURE 4.1 Translation of an object at (x, y, z) along a glide plane

same as that of the next person, in order to explain or discuss things in three dimensions we need to find a unified starting point. To do this, it becomes necessary to convey three-dimensional messages in two dimensions (namely, paper). Pictorially, crystallographers use space-group diagrams.

A *space-group* diagram is a *two-dimensional* projection of a *three-dimensional* unit cell. Figure 4.2 shows a three-dimensional view of a unit cell in solid black lines, with the dotted boxes dividing the unit cell along each of the axes into two equal halves. In the lower back left corner of the unit cell there is an object (in blue, which represents any unit cell content, atom, ion, or molecule). The positional coordinates of this object are (a, b, c), which can be described as a value of a along x, a value of b along y, and a value of c along z.

If we can imagine looking downwards on that unit cell in Fig. 4.2, we would then be able to draw a plan view of that unit cell. In the space-group diagram in Fig. 4.3, this plan view is known as the **general positions diagram**. It typically contains objects in the equivalent positions within the unit cell.

Based on Fig. 4.2, the view would be looking down the c-axis, however in the diagrams from the *International Tables* these axes may be changed and are clearly labelled in the diagrams.

Also on the same page for each space group are the **symmetry-elements diagrams.** These diagrams provide information on the symmetry elements that occur in each space group. The general positions diagrams and the symmetry-elements diagrams complement each other.

Figure 4.3 is an example page taken from the fifth edition of the *International Tables for Crystallography* (labels in blue are for clarification). At the very top of each page (from left to right) are the Hermann–Mauguin notation and the consecutive numbering of the space

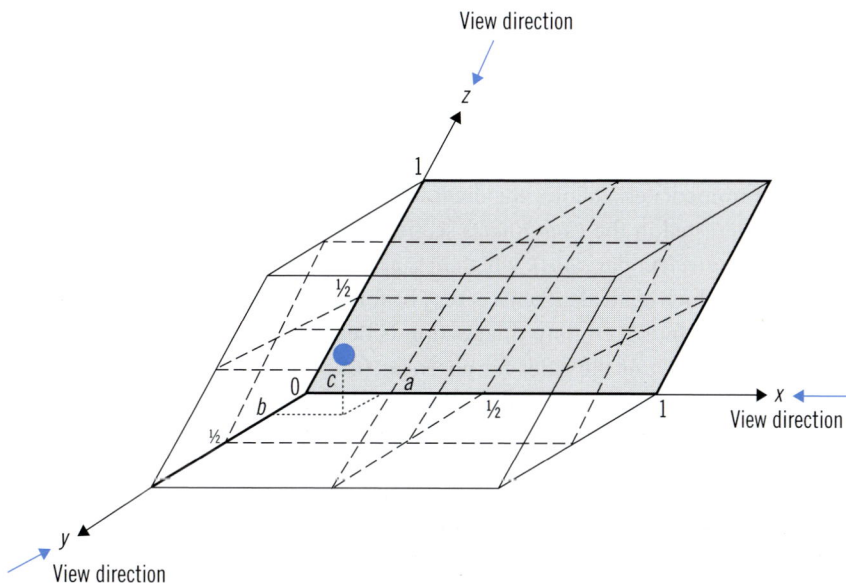

FIGURE 4.2 The three-dimensional viewing directions for space-group diagrams

Hermann - Mauguin notation Consecutive Numbering	Schöenflies symbol	Laue class	Crystal System and Space group
$P1$	C_1^1	1	Triclinic
No. 1	$P1$		Patterson symmetry $P\,1$

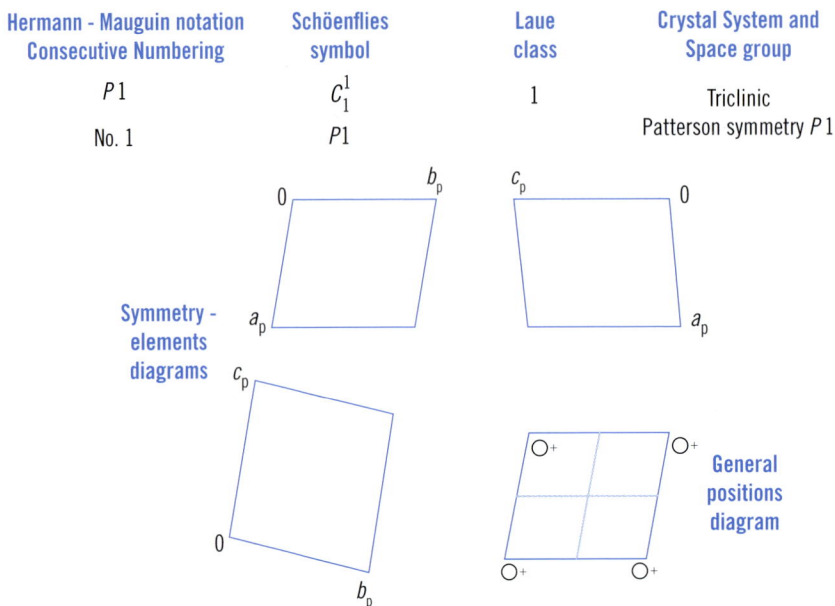

FIGURE 4.3 A page (space-group diagrams) from the *International Tables for Crystallography*, labels in blue.

Reproduced from: Hahn, T. (ed.) (2002). *International Tables for Crystallography: Volume A (Space-group Symmetry)*, 5th edn. Kluwer Academic.

groups followed by the Schöenflies symbol. These are followed by the *Laue* class assignment for the space group, and in the top right corner of the page are the abbreviated crystal system and the space group.

This information is followed by line diagrams of the symmetry-elements diagrams, with the axes and view directions marked and the general position diagram in the lower right corner. The *origin*, the asymmetric unit, and the number of symmetry operations are noted in the lower third of the page.

The page that immediately follows the diagrams (Fig. 4.4) lists several useful details. We are mainly interested in the *Coordinates* section (in the fourth edition, this is listed as *Coordinates of equivalent positions*), which lists all the possible positional coordinates within the lattice at which equivalent objects may be found and the section marked *Reflection conditions* (in the fourth edition, this section is headed *Conditions limiting possible reflections*), which is a list of the rules for absences (of reflections), which should correspond with the collected diffraction data. Absences in diffraction data are explained further in Chapter 5.

In the *International Tables for Crystallography* (fifth edition), the three symmetry-elements diagrams are views along each of the three axes. In each of the symmetry-elements diagrams, it is necessary to be able to convey the type of symmetry operator as well as the axis along which this operation is relevant. For this purpose, graphical symbols representing the symmetry operations are used.

4.2.3 Graphical symbols in space groups

Graphical symbols provide, pictorially, a point of reference as to the type of symmetry operator and its location within a unit cell.

As the symmetry-elements diagram can be viewed from any of the three axes, there are also symbols denoting the type of symmetry element from a plane view or a parallel view of the unit cell. Earlier, in Chapter 3, graphical symbols were introduced alongside the descriptions of the symmetry operations. These symbols are used and applied in the symmetry-elements diagrams.

Some of the more commonly used symbols are listed in Table 4.2.

CONTINUED   No. 1 $P1$

Generators selected   (1): $i(1, 0, 0)$: $i(0, 1, 0)$: $i(0, 0, 1)$

Positions
Multiplicity   **Coordinates**   Reflection conditions
Wyckoff letter
Size symmetry   General:

1   ii   I   (1) x, y, z no conditions

Symmetry of special projections
Along [001] $p1$ Along [100] $p1$ Along [010] $p1$
$a' = a_p$, $b' = b_p$ $a' = b_p$, $b' = c_p$ $a' = c_p$, $b' = a_p$
Origin at $x, 0, 0, z$ Origin at $x, 0, 0$ Origin at $0, y, 0$

Maximal non-isomorphic subgroups
I   none
IIa   none
IIb   none

Maximal isomorphic subgroups of lowest index
IIc   [2] $P1$ ($a = 2a$ or $b' = 2b$ or $c' = 2c$ or $b' = b + c$, $c' = -b + c$ or $a' = a - c$, $c' = a + c$ or $a' = a + b$, $b' = -a + b$
         or $a' = b + c$, $b' = a + c$, $c' = a + b$) (1)

Minimal non-isomorphic subgroups
I   [2] $P1$ (2r, [2] $P2$ (3); [2] $P2_1$ (4); [2] $C2$ (5); [2] Pm (6); [2] Pc (7); [2] Cm (8); [2] Cc (9); [3] $P3$ (143); [3] $P3_1$ (144);
         [3] $P3_2$ (145); [3] $R3$ (146)
II   none

FIGURE 4.4 The following page from the *International Tables for Crystallography*.
Reproduced from: Hahn, T. (ed.) (2002). *International Tables for Crystallography: Volume A (Space-group Symmetry)*,
5th edn. Kluwer Academic.

TABLE 4.2 Summary of common symbols and notations used in crystallography

Non-translational symmetry elements	Notation	Symbol
Inversion	$\bar{1}$ (bar one)	•
Mirror	m	⌐ —
Rotations		
Diad	2-fold	◆ ✦
Triad	3-fold	▲ ▲

(*Continued*)

TABLE 4.2 (*Continued*)

Non-translational symmetry elements	Notation	Symbol
Tetrad	4-fold	◆ ✦
Hexad	6-fold	● ✦ ♦
Rotation-inversions		
Diad	2-fold	◊
Triad	3-fold	△
Tetrad	4-fold	◇
Hexad	6-fold	⬡
Translational symmetry elements	**Notation**	**Symbol**
Screw axes		
Diad	2_1	↯
Triad	$3_2, 3_1$	▲▲
Tetrad	$4_1, 4_2, 4_3$	✦◆✦
Hexad	$6_1 \ 6_2 \ 6_3 \ 6_4 \ 6_5$	✦◆♦ ◆✦
Glide planes	*a, b, c, n*	-----

SELF-TEST QUESTIONS

1. Explain what the following Hermann–Mauguin notations represent:

   (a) *P* 2 2 2;

   (b) *C* 2;

   (c) *C* m;

   (d) *I b a* 2.

2. On a three-dimensional lattice diagram, mark the following glide planes:

   (a) *a* glide perpendicular to *b*;

   (b) *b* glide perpendicular to *c*;

   (c) *b* glide perpendicular to *a*.

3. With the aid of diagrams (and symbols), describe the movement of an object undergoing a '*b* glide perpendicular to *a*'.

4.3 DECIPHERING SPACE-GROUP DIAGRAMS

To be able to 'read' a space-group diagram, there are some conventions that first need to be identified. In the general position diagram, a hollow circle (○) is commonly used to

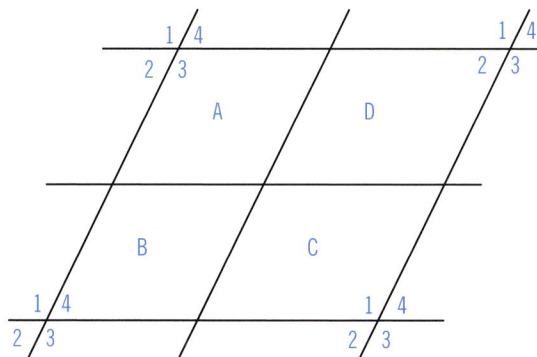

FIGURE 4.5 Quadrant labels

represent an object. A circle with a comma inside it (⊙) denotes the mirror image of that same object. A plus sign (+) means that the object is above the plane of the lattice while a minus sign (−) represents an object below the plane of the lattice.

The objects within the general position diagram are subjected to the symmetry operations in the symmetry-elements diagrams. However each 'object' only undergoes *one* symmetry operation at any one time. In deciphering the space-group diagrams from the *International Tables*, we are mainly concerned with the general position diagram (at the lower right corner) and its equivalent symmetry-elements diagram (at the top left corner).

In the following sections, we shall examine some space-group diagrams of some of the more common space groups. For convenience of description and reference, the quadrants are labelled within the general position diagram.

The labels in Fig. 4.5 will be used in the following space-group descriptions and are used here only as a point of reference. They do not appear in the *International Tables for Crystallography* and are not a convention.

4.3.1 Triclinic

Owing to its lack of lattice symmetry, there are only two possible space groups for the triclinic lattice. They are $P\,1$ (*P* one) and $P\,\bar{1}$ (pronounced *P* bar one).

Triclinic P1

The triclinic $P\,1$ space group (see Fig. 4.6) contains no symmetry elements; hence there are no symbols in the symmetry-elements diagram. Examining the general position diagram (lower right) in conjunction with the relevant space-group diagram (top left), we find that each lattice contains only one object. The coordinates of this object are given as *x*, *y*, and *z*.

Triclinic P$\bar{1}$

The triclinic $P\,\bar{1}$ space group (see Fig. 4.7) contains one inversion centre. In the symmetry-elements diagram, this is represented by the circle in the middle of the diagram. In the general position diagram, we find the object in quadrant A3 at coordinates (*x, y, z*). In undergoing an inversion, the object becomes its mirror image in quadrant C1 at (−*x*, −*y*, −*z*). Therefore, the coordinates of the equivalent positions are (*x, y, z*) and (−*x*, −*y*, −*z*).

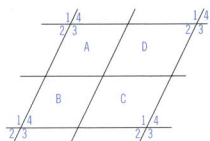

$P\,1$ C_1^1 1   Triclinic

No. 1 $P1$ Patterson symmetry $P\,1$

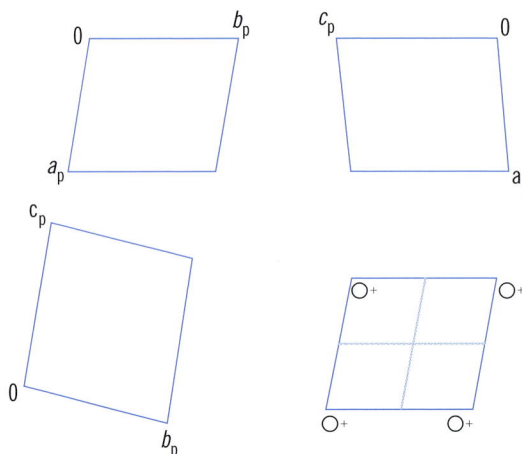

Drawings for type II cell. Proper cell reductions gives either
a type I (α, β, γ acute) or a type II (α, β, γ non-acute) cell

Origin arbitrary
Asymmetric unit $0 \leq x \leq 1;\quad 0 \leq y \leq 1;\quad 0 \leq z \leq 1$
Symmetry operations   (1) 1

FIGURE 4.6 Space-group diagram for triclinic $P\,1$

Reproduced from: Hahn, T. (ed.) (2002). *International Tables for Crystallography: Volume A (Space-group Symmetry)*,
5th edn. Kluwer Academic.

This lattice content is then repeated in the adjacent unit cells, giving rise to the other objects
located around the diagram in quadrants B3, C3, and D3 and the inverted objects in A1, B1,
and D1. This also results in inversion centres at the edges of the lattice and along the axes.

4.3.2 Monoclinic

Monoclinic P m

The monoclinic space group $P\,1m\,1$, denotes a primitive lattice with a mirror plane
perpendicular to the b-axis. In the symmetry-elements diagrams, the bold line represents
the mirror plane that is perpendicular to the horizontal b-axis. In Fig. 4.8, we are comparing the general position diagram (lower right) and the symmetry-elements diagram (top
left). The view of the symmetry-elements diagram in this case is looking down the b-axis,
in essence looking on to the mirror plane. The symbol (-⊕) denotes two objects in the same
vertical position: an object (○+) above the plane and its mirror image (-⊙) below the plane.

For ease of visualization, we refer briefly to the same space group from the earlier fourth
edition of the *International Tables* in Fig. 4.9. Here we have the general position diagram on
the right and the symmetry-elements diagram on the left. Both are views down the c-axis.

In the general position diagram, if we start with the object in A3 at (x, y, z), the object
is reflected across the first mirror plane resulting in the object at A2. As this object is the

$P\,\bar{1}$ C_i^1 $\bar{1}$ Triclinic

No. 2 $P\,\bar{1}$ Patterson symmetry $P\,\bar{1}$

symmetry

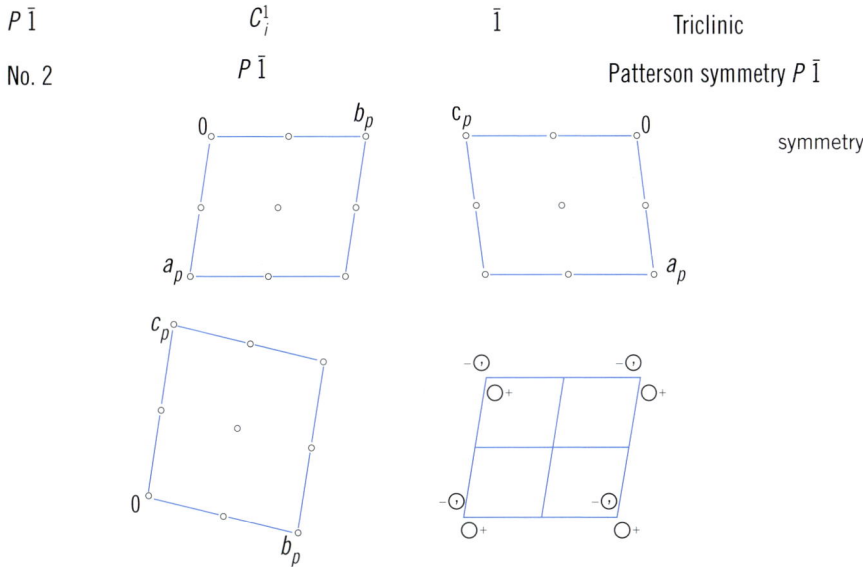

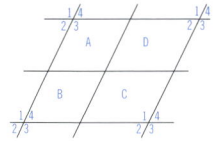

Drawings for type II cell. Proper cell reduction gives either
a type I (α, β, γ acute) or a type II (α, β, γ non-acute) cell

Origin at $\bar{1}$
Asymmetric unit $0 \leq x \leq \frac{1}{2}; 0 \leq y \leq 1; 0 \leq z \leq 1$
Symmetry operations   (1) 1 (2) $\bar{1}$ 0, 0, 0

FIGURE 4.7 Space-group diagram for triclinic $P\,\bar{1}$
Reproduced from: Hahn, T. (ed.) (2002). *International Tables for Crystallography: Volume A (Space-group Symmetry)*,
5th edn. Kluwer Academic.

direct mirror image, the coordinates are $(x, -y, z)$. Therefore, the coordinates of equivalent positions are (x, y, z) and $(x, -y, z)$.

Equally, the object in A3 may be reflected across the central mirror plane, giving rise to the object in quadrant D2, which also has the coordinates $(x, -y, z)$. The equivalent objects are also found in the adjacent unit cells, resulting in the objects found in B2 and B3, C2 and C3.

If we look again at Fig 4.8, we can identify with the location of each of the objects corresponding to the view of the equivalent positions diagram, looking down the *b*-axis.

Monoclinic P c

The monoclinic space group $P\,1c\,1$ (see Fig. 4.10) denotes a primitive lattice with a *c*-glide perpendicular to the *b*-axis. (A glide is a translation followed by a reflection.) In the symmetry-elements diagrams, the dotted lines represent the plane view of the *c-glide* plane lying perpendicular to the horizontal *b*-axis. This was visually explained in Section 4.2.1.

In the general position diagram (looking down the *b*-axis), if we start with the object in A3 at (x, y, z), the object is translated along the *c*-axis by half the unit cell and then reflected across the glide plane, resulting in the object at B4 (a mirror image of the original object translated upwards by half a unit cell). The coordinates of this 'new' object are $(x, -y, z + \frac{1}{2})$. Therefore, the coordinates of equivalent positions are (x, y, z) and $(x, -y, z + \frac{1}{2})$.

Pm

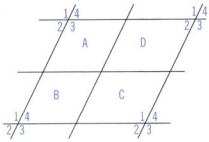

No. 6
Unique axis b

C_s^1

$P1m1$

m

Monoclinic
Patterson symmetry $P12/m1$

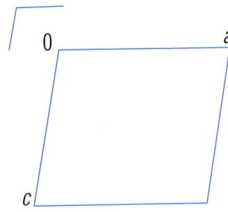

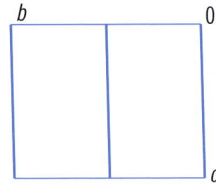

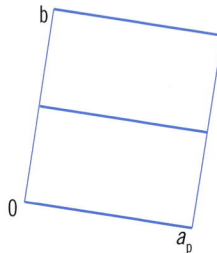

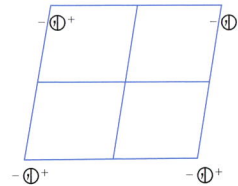

Origin on mirror plane m
Asymmetric unit $0 \leq x \leq 1; 0 \leq y \leq \frac{1}{2}; 0 \leq z \leq 1$
Symmetry operations   (1) 1   (2) $m\ x, 0, z$

FIGURE 4.8 Space-group diagram for monoclinic Pm

Reproduced from: Hahn, T. (ed.) (2002). *International Tables for Crystallography: Volume A (Space-group Symmetry)*, 5th edn. Kluwer Academic.

Monoclinic m

$P1m1$

No. 6

Pm
C_s^1

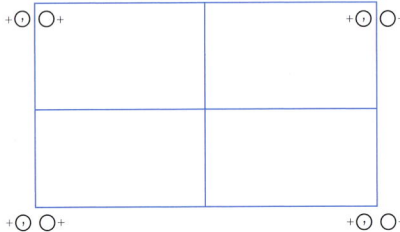

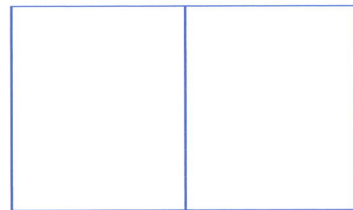

Number of positions,
Wyckoff notation,
and point symmetry

Origin on plane m; unique axis b
coordinates of equivalent positions

2nd setting
Conditions limiting
possible reflections

•	c	1
1	b	m
1	a	m

x, y, z
$x, -y, z$

General:
hkl:
h0l: No conditions
0k0:

Special:
No conditions

Symmetry of special projections

(001) $p1m$; $a' = a$, $b' = b$
(100) $pm1$; $b' = b$, $c' = c$
(001) p; $c' = c$, $a' = a$

FIGURE 4.9 Space-group diagram for monoclinic Pm (fourth edn)

Reproduced from: Hahn, T. (ed.) (1995). *International Tables for Crystallography: Volume A (Space-group Symmetry)*, 4th edn. Kluwer Academic.

Pc $\qquad$ C_s^2 $\qquad$ m $\qquad$ Monoclinic

No. 7 $\qquad$ $P1\,c1$ $\qquad$ Patterson symmetry $P12/m1$

Unique axis b, cell choice 1

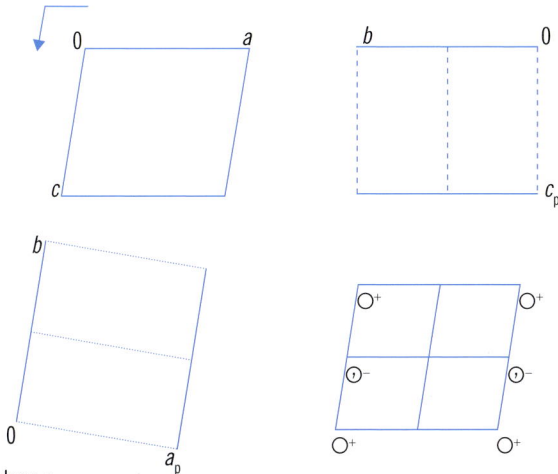

Origin on glide plane c

Asymmetric unit $\qquad$ $0 \leq x \leq 1; 0 \leq y \leq \frac{1}{2}; 0 \leq z \leq 1$

Symmetry operations $\qquad$ (1) 1 (2) c x, 0, z

FIGURE 4.10 Space-group diagram for monoclinic Pc

Reproduced from: Hahn, T. (ed.) (2002). International Tables for Crystallography: Volume A (Space-group Symmetry), 5th edn. Kluwer Academic.

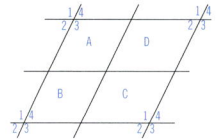

The content of this lattice is also identical in the adjacent unit cells, which result in the object found in the quadrants B3, C3 and C4, and D3.

Monoclinic P 2_1/m

The space group $P 1\, 2_1/m\, 1$ (see Fig. 4.11) denotes a primitive lattice with a 2_1 screw axis along the b-axis and a mirror plane perpendicular to the b-axis. (A 2_1 screw axis is a 180° rotation followed by a translation upwards of half the unit cell.)

In the symmetry-elements diagram, the 2_1 screw axis is denoted by the half arrows along the b-axis while the mirror planes are the bold lines perpendicular to the b-axis.

Again for ease of visualization, we refer briefly to the same space group from the fourth edition of the *International Tables* in Fig. 4.12. Here, we have the equivalent positions diagram on the right and the symmetry-elements diagram on the left. Both are views down the c-axis.

If we consider first the 2_1 screw axis, starting with the object at (x, y, z) in the quadrant A3, undergoing a 180° rotation followed by a translation half the length of the b-axis, the object would now be located in the D1 quadrant slightly below the lattice at $(-x, \frac{1}{2} + y, -z)$.

If we are then to consider the mirror planes, again starting with the object at (x, y, z) in A3, reflected across the mirror plane perpendicular to the b-axis, we would find the resulting object in mirror image again in A3 at $(x, \frac{1}{2} - y, z)$.

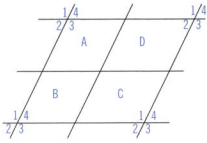

$P2_1/m$ C_{2h}^2 $2/m$ Monoclinic

No. 11 $P12_1/m1$ Patterson symmetry $P12/m1$

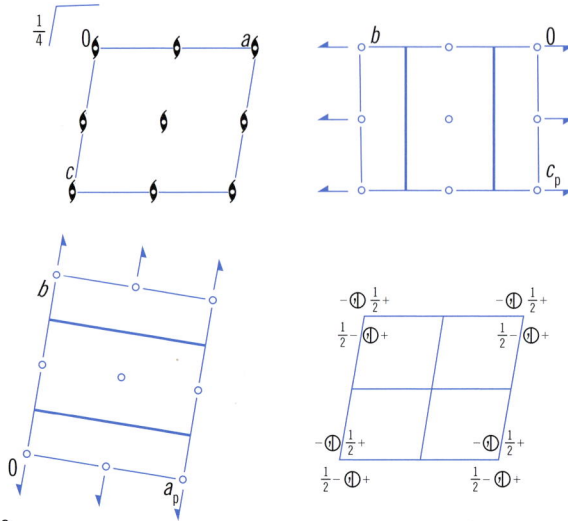

Unique axis b

Origin at $\bar{1}$ on 2_1

Asymmetric unit $0 \le x \le 1; 0 \le y \le \frac{1}{4}; 0 \le z \le 1$

Symmetry operations (1) 1   (2) $2(0, \frac{1}{2}, 0)$ 0, y, 0   (3) $\bar{1}$ 0, 0, 0   (4) m $x, \frac{1}{4}, z$

FIGURE 4.11 Space-group diagram for monoclinic P 2_1/m

Reproduced from: Hahn, T. (ed.) (2002). *International Tables for Crystallography: Volume A (Space-group Symmetry)*, 5th edn. Kluwer Academic.

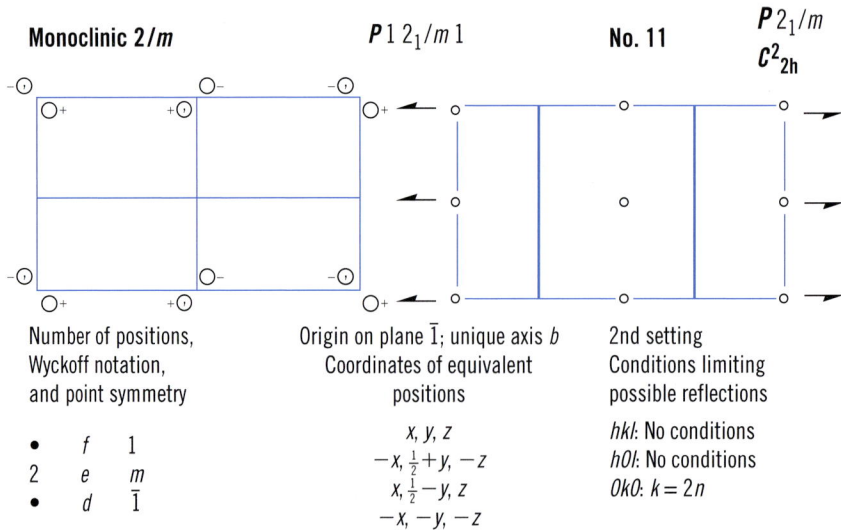

Monoclinic 2/m $P1\,2_1/m\,1$ No. 11 $P\,2_1/m$
C^2_{2h}

Number of positions, Wyckoff notation, and point symmetry	Origin on plane $\bar{1}$; unique axis b Coordinates of equivalent positions	2nd setting Conditions limiting possible reflections
• f 1	x, y, z	*hkl*: No conditions
2 e m	$-x, \frac{1}{2}+y, -z$	*h0l*: No conditions
• d $\bar{1}$	$x, \frac{1}{2}-y, z$	*0k0*: $k = 2n$
	$-x, -y, -z$	

FIGURE 4.12 Space-group diagram for monoclinic P 2_1/m *(fourth edition)*

Reproduced from: Hahn, T. (ed.) (1995). *International Tables for Crystallography: Volume A (Space-group Symmetry)*, 4th edn. Kluwer Academic.

The object at D1 $(-x, \frac{1}{2} + y, -z)$ will also be reflected across the mirror plane resulting in a mirror image located at A1 at $(-x, -y, -z)$. Therefore, all the coordinates of equivalent positions are (x, y, z), $(x, \frac{1}{2} - y, z)$, $(-x, \frac{1}{2} + y, -z)$, and $(-x, -y, -z)$.

Referring again to Fig 4.11, we can also identify with the location of each of the objects corresponding to the view of the equivalent positions diagram, looking down the b-axis, down the screw axes.

SELF-TEST QUESTIONS

1. Describe the symmetry element or elements you would expect to find in the orthorhombic space group $P\,2_1\,2_1\,2$ and $P\,2_1\,2_1\,2_1$. Briefly explain how the two differ.

2. A platinum (Pt) complex crystallizes in the monoclinic space group $P\,2_1/m$. One of the Pt atoms is located at $(0.3, 0.3, -0.6)$. Based on the coordinates of equivalent positions, where would you expect to find the other Pt atoms?

3. For the orthorhombic space group $P\,2_1\,2_1\,2_1$, fill in the equivalent positions diagram with symmetry-related objects based on the symmetry-elements diagram provided.

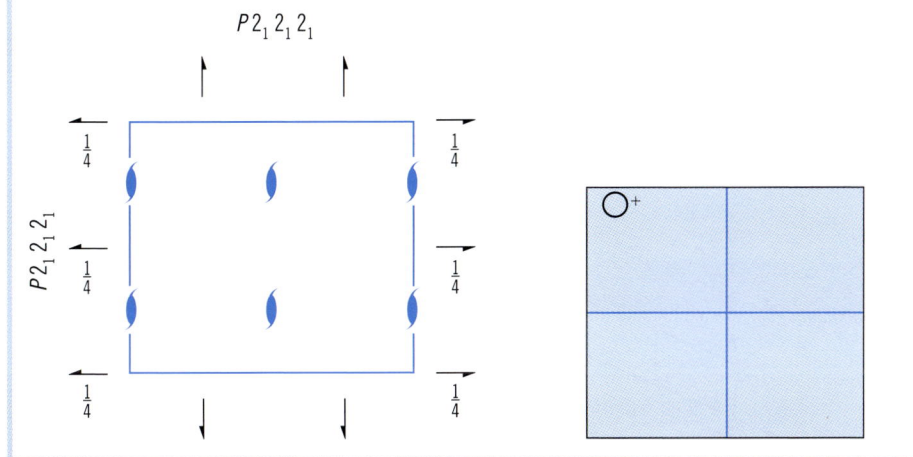

$P\,2_1\,2_1\,2_1$

4.4 SPECIAL POSITIONS IN SPACE GROUPS

So far we have considered mainly the consequences of symmetry in relation to the packing of the molecules in the solid state, that is, how each molecule is related to the next within the crystal lattice. However, there are occasions when there is symmetry within the molecule under study. In cases like these it is often common to find that the symmetry within the molecule contributes to the packing of the crystal lattice, thereby contributing in some way to the symmetry of the unit cell.

In essence, an atom (or atoms) is considered to be in a special position, when its symmetry equivalent point maps onto itself by another symmetry operation.

(a)

(b)

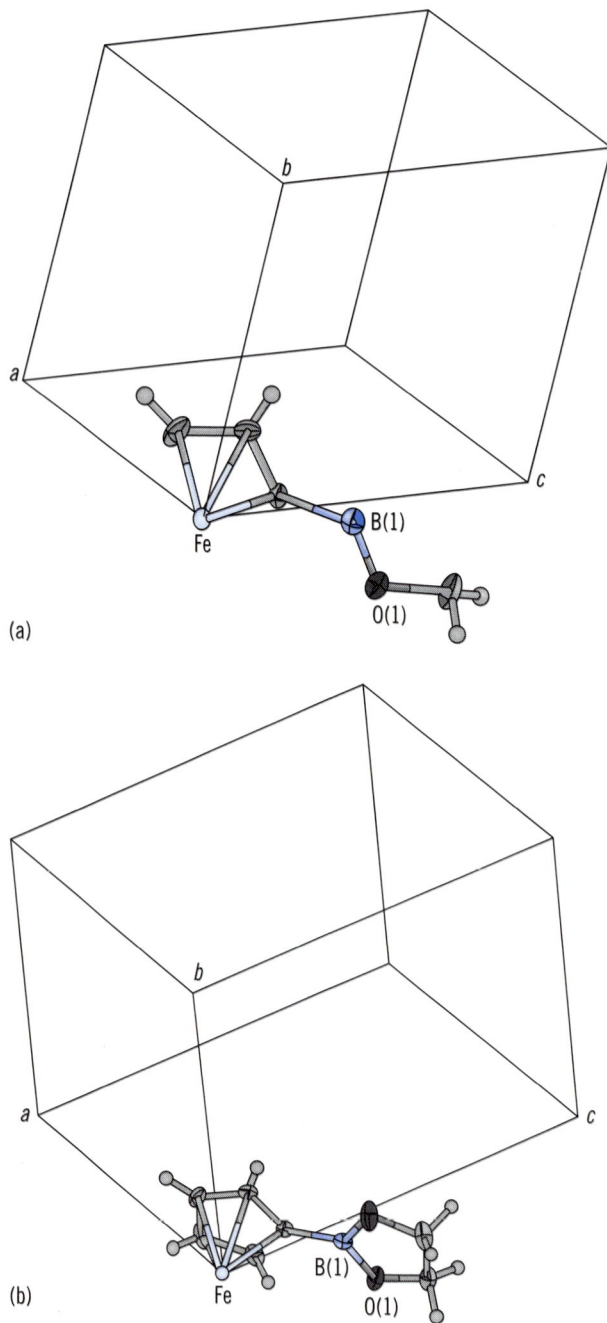

FIGURE 4.13 A ferrocene derivative with atoms in special positions

In Fig. 4.13, the molecules crystallize in the monoclinic space group $C\ 2/m$. This means that within the C-centred monoclinic lattice, there is a mirror plane and a two-fold rotation.

As a result, the asymmetric unit, in Fig 4.13(a) consists of a quarter of the molecule with the atoms Fe and B(1) in special positions, while another quarter of the molecule maps

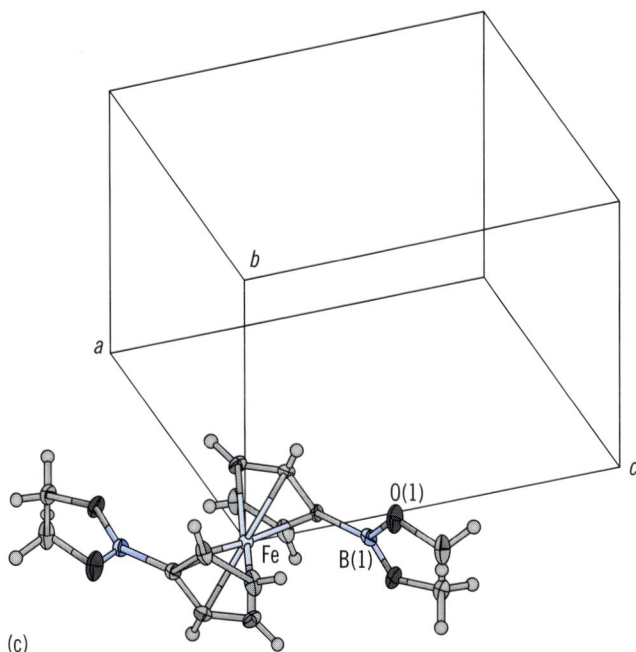

(c)

FIGURE 4.13 Continued

onto those same atoms, as shown in Fig 4.13(b). The other half of the molecule again maps onto the Fe atom; the complete molecule is shown in Fig 4.13(c).

4.5 CENTROSYMMETRIC AND NON-CENTROSYMMETRIC SPACE GROUPS

All of the 230 known space groups can be broadly divided into two large groups, centrosymmetric and non-centrosymmetric space groups.

In essence, a centrosymmetric space group is, like the name suggests, one that contains a centre of symmetry. Usually, this means a space group that has an inversion centre, as it is through this inversion centre that a unit cell in a centrosymmetric space group would contain an object at (x, y, z) and at $(-x, -y, -z)$. The most common example is the triclinic $P\bar{1}$.

A non-centrosymmetric space group is one that does not contain a discernable centre of symmetry. These are generally the space groups that do not contain an inversion centre. The non-centrosymmetric space groups are again subdivided into chiral and achiral (polar) space groups.

Chiral space groups are also known as Sohnke groups. There are 65 space groups that contain only rotation or screw axes, and it is only in one of these space groups that a chiral molecule may crystallize. As shown in Fig. 4.14, chiral space groups, however, are not mutually exclusive, so while a chiral molecule may only crystallize in a chiral space group, non-chiral molecules may still crystallize in a chiral space group. The rest of the non-centrosymmetric space groups are made up of achiral space groups.

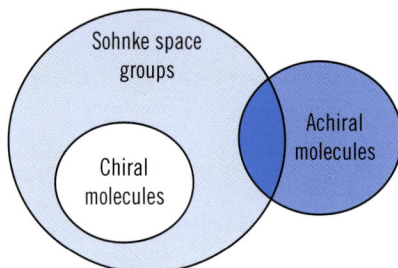

FIGURE 4.14 A Venn diagram representing the association of Sohnke space groups with chiral and achiral molecules

4.6 SYMMETRY IN MACROMOLECULAR (PROTEIN) CRYSTALS

Proteins or macromolecules are, in essence, polymers built from a range of 20 amino acids. Each of these amino acids is chiral with two configurations, a right and a left; conventionally they are termed D- (dextrorotatory) and L- (levorotatory). A chiral molecule is a molecule that cannot be superimposed on to its own mirror image. For example, alanine, which is shown in Fig. 4.15.

As each of these amino acid building blocks is chiral, the resulting protein is also a chiral molecule. This also means that when the protein molecule crystallizes it cannot contain a mirror plane or an inversion centre. If we refer to Section 4.5, we remember that all chiral molecules will only crystallize in one of the 65 chiral space groups. Hence, out of the known 230 space groups, protein crystals can only crystallize in one of the 65 chiral space groups.

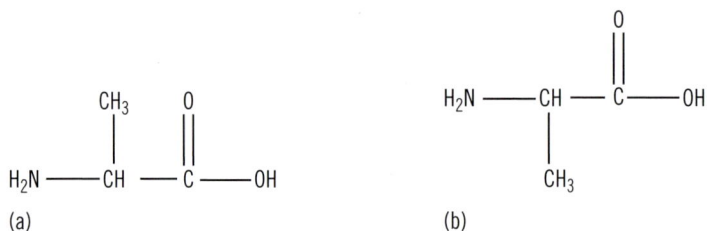

FIGURE 4.15 (a) L-alanine, (b) D-alanine

SELF-TEST QUESTIONS

1. Briefly describe 'special positions' in the context of a molecule in a crystal lattice.

2. What are centrosymmetric and non-centrosymmetric space groups? Give examples of each.

3. What types of molecules *have* to crystallize in a Sohnke space group? Give an example and explain why.

◻ CHAPTER SUMMARY

1. In crystallography, there are 230 unique arrangements of objects in three-dimensional space. These are the 230 known space groups.

2. The Hermann–Mauguin notation represents the lattice type and the symmetry elements.

3. A space-group diagram is a two-dimensional projection of a three-dimensional unit cell.

4. A space-group diagram consists of the symmetry-elements diagrams, a general position diagram, and all other information pertaining to a particular space group.

5. Graphical symbols are used to represent symmetry elements within the symmetry-elements diagrams.

6. An atom (or atoms) is considered to be in a special position when its symmetry equivalent points, maps onto itself by another symmetry operation.

7. All 230 space groups can be divided into centrosymmetric and non-centrosymmetric space groups. Non-centrosymmetric space groups can be further divided into chiral and achiral space groups.

8. Chiral molecules can only crystallize into one of the 65 chiral (Sohnke) space groups.

9. As all proteins are chiral molecules, they can only crystallize into one of the 65 chiral space groups.

📖 FURTHER READING

Giacovazzo, C., Monaco, H. L., Artioli, G., *et al.* (eds.) (2002). *Fundamentals of Crystallography,* 2nd edn. IUCr Texts on Crystallography. Oxford University Press, New York.

Hahn, T. (ed.) (1995). *International Tables for Crystallography: Volume A (Space-group Symmetry),* 4th edn. Kluwer Academic, Dordrecht.

Hahn, T. (ed.) (2002). *International Tables for Crystallography: Volume A (Space-group Symmetry),* 5th edn. Kluwer Academic, Dordrecht.

Hahn, T. (ed.) (2002). *International Tables for Crystallography: Brief Teaching Edition of Volume A (Space-group Symmetry),* 5th edn. Kluwer Academic, Dordrecht.

🐾 LINKS

Online learning tools
Experiment with space groups and symmetry: http://escher.epfl.ch/cse

SYSTEMATIC ABSENCES IN CRYSTAL DATA

5

By the end of this chapter you should be able to:

- Understand how absences in crystal data relate to lattice types and symmetry elements;
- Recognize how intensities in crystal data are associated with structure factors;
- Gain an appreciation for atomic scattering factors and its influence on structure factors;
- Identify the rules governing general and systematic absences.

5.1 INTRODUCTION

When a diffraction pattern is collected either on photographic film or on an X-ray diffractometer, each 'frame' of data (see Fig. 5.1) contains reflections (spots) of differing intensities, while at certain points on the frame of data some reflections are missing or *absent*.

Some of these absent reflections have intensities close to zero because only very few electrons in the crystal structure are contributing to diffraction from the associated plane, while other reflections are precisely zero because of the destructive interference of the incident X-rays. These absences are influenced by the positions of symmetry-related atoms or molecules within the crystal structure.

In this chapter, we will examine how *absences* (absent reflections) can be used to determine the space group of a crystal lattice, relating directly to the various lattice types and the different translational symmetry elements that can occur within a crystal lattice.

During the course of a data collection, multiple *frames* of data are collected, in order to accumulate sufficient data to cover the reciprocal space of the entire crystal lattice. Within each frame of data, each reflection is related to a specific Miller plane (h, k, l) with a specific value of 'observed' intensity, I_{hkl}^{obs} used to define it. An *absence* occurs when $I_{hkl}^{obs} = 0$.

Once the complete diffraction data are collected, the data are processed. The processing of the diffraction pattern into a useable format is also known as *data reduction*. The integration of data occurs as part of the data-reduction process. During integration, the diffraction

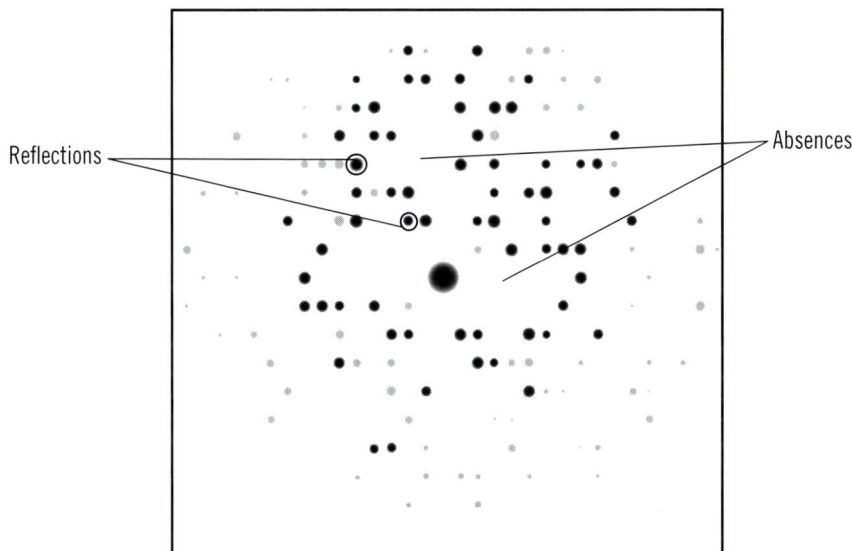

FIGURE 5.1 A frame of diffraction data

from each Miller plane within the single crystal lattice is integrated across the multiple frames of diffraction data. This ensures that each collected diffraction spot is linked to a Miller index and has an associated value of intensity. Each integrated value of intensity is proportional to the square of the observed **_structure factor_** (F_{hkl}^{obs}) for the associated Miller index;

$$I_{hkl}^{corr} \propto (F_{hkl}^{obs})^2.$$

After integration, scaling, and various corrections for background (Lorentz, polarization, and absorption corrections) are applied, culminating in the output of a text-based computer file of diffraction data representative of the single crystal. This is known as the *.hkl* file. The *.hkl* file (see Fig. 5.2) contains Miller indices (*hkl* values) in the first three columns, followed by a numerical value of intensity (*I*) and a value of standard deviation (σ).

The computer file name extensions used in this text, such as *.hkl*, *.res*, and *.ins*, refer to those used by the SHELX suite of programs.

5.2 STRUCTURE FACTORS

Section learning outcomes

To be able to:

- Understand the contributions of atomic scattering factors to structure factors;
- Calculate structure factors and derive calculated intensities.

From experimental diffraction data, we are able to obtain 'observed and corrected' intensities I_{hkl}^{corr}, directly related to the square of the 'observed' structure factors $(F_{hkl}^{obs})^2$. The diffraction

h	k	l	I	σ	
−2	2	−2	124.99	6.26	1
−2	2	−1	971.57	38.65	1
−2	2	7	1176.49	100.16	1
−2	2	8	0.59	0.89	1
−1	−7	10	443.52	26.96	1
−1	−7	11	5.52	4.61	1
−1	−6	4	696.96	40.66	1
−1	−5	−8	5.52	2.91	1
−1	−1	11	44.36	3.73	1
−1	0	−2	1043.94	69.79	1
−1	0	−2	1622.48	66.06	1
−1	1	8	85.56	5.36	1
−1	2	10	1.59	1.29	1
−1	3	−1	530.38	21.65	1
−1	3	14	451.99	35.72	1
−1	4	−12	47.33	6.74	1
0	−13	5	113.42	13.21	1
0	−12	−7	298.94	23.86	1
0	−10	10	70.73	9.42	1
0	−9	−11	0.10	6.68	1
0	−8	−7	312.58	16.27	1
0	−7	−3	1.21	1.10	1
0	−6	−1	159.26	8.08	1
0	−6	−1	138.30	6.35	1
0	−5	3	758.45	38.01	1
0	−4	0	1710.65	82.72	1
0	−2	9	1.80	1.90	1
0	−1	−3	649.74	37.22	1
0	0	4	673.40	36.33	1
0	0	4	1202.01	52.00	1
0	1	6	509.40	27.08	1
0	2	−7	900.60	55.22	1
0	4	0	1443.24	73.70	1
0	5	3	776.18	37.33	1
1	−7	0	1147.18	47.42	1

hkl values

σ, standard deviation

I, intensity

FIGURE 5.2 Part of an *.hkl* file

pattern produced is a result of the total scattering from all unit cells and represents the average content of a single unit cell. Structure factors are, in essence, a mathematical representation of the interaction between molecules in a crystal lattice with X-rays and are influenced by atomic scattering factors.

TABLE 5.1 Atomic scattering factors for some elements

Prince, E. (ed.) (2004). *International Tables for Crystallography Volume C: Mathematical, Physical and Chemical Tables*, 3rd edn. Kluwer Academic.

$\sin\theta/\lambda$	H	C	O	O⁻	Na	Al	Ca	Sn	W
0	1	5	8	9	11	13	20	50	74
0.1	0.811	5.13	7.245	7.836	11.0	11.23	17.33	46.36	69.78
0.2	0.481	3.58	5.623	5.623	9.76	9.16	14.32	40.30	62.52
0.5	0.071	1.69	2.338	2.313	4.29	5.69	8.26	29.10	43.70
1.0	0.007	1.11	1.377	1.376	1.78	2.32	5.19	16.38	25.58

5.2.1 Atomic scattering factors

The electrons in atoms are able to interact with and subsequently scatter X-rays. Atomic scattering factors, f_n, also known as atomic form factors, are the amplitude measure of X-ray waves scattered from an atom.

As X-rays are scattered by the electrons within an atom, the atomic scattering factor is dependent on the electron density exhibited by the atom or the atomic number within an atom. Table 5.1 lists the atomic scattering factors for some of the elements and how they are related to the Bragg angle. Referring to Table 5.1, we can see that the atomic scattering factor of hydrogen (one electron) is significantly different from that of tin (50 electrons).

From Table 5.1, we can also notice that in the heavy elements, the reduction at high Bragg angles occurs more slowly. This indirectly allows the heavy elements to dominate the structure factors, making it more difficult to locate the lighter atoms within the diffraction map.

The atomic scattering factor is a function of the Bragg angle. For example, if we refer to Fig. 5.3, at low Bragg angles (e.g., when $\sin\theta/\lambda = 0$), the atomic scattering factor, f_n, is directly proportional to the atomic number. In contrast, this value tails off at higher angles, as at these high angles not all of the electrons within an atom are scattering in phase.

5.2.2. Calculating structure factors and intensities

The calculated structure factors F_{hkl}^{calc} relating to a given Miller index (h, k, l) for a crystal containing N atoms is given by the following equation:

$$F_{hkl}^{calc} = \sum_{1}^{N} f_n \cos 2\pi(hx_n + ky_n + lz_n) + i\sum_{1}^{N} f_n \sin 2\pi(hx_n + ky_n + lz_n), \qquad (5.1)$$

where x_n, y_n, and z_n denote the fractional coordinates of the atoms in the structure while f_n denotes the atomic scattering factor for atom type of n.

i is a complex number, $\sqrt{-1}$

⊕ Learning aid – breaking down the equation

In reality what seems to be a complex equation can actually be broken down into several familiar sections. Within the brackets, $h\,k\,l$ and $x\,y\,z$ are familiar terms relating to specific Miller indices and the fractional coordinates of an atom, and f_n is the atomic scattering factor.

The atomic scattering factor is influenced by both the scattering from valence electrons, $f_{valence}$, and core electrons, f_{core}, given by:

$$f_n = f_{valence} + f_{core}.$$

Scattering of the valence electrons occurs most efficiently at low Bragg angles, while at high Bragg angles, the scattering of the core electrons is more significant. As the heavy elements contain more core electrons, the atomic scattering factor tends to diminish more slowly at higher Bragg angles.

Plot of sinθ/λ vs. f_0

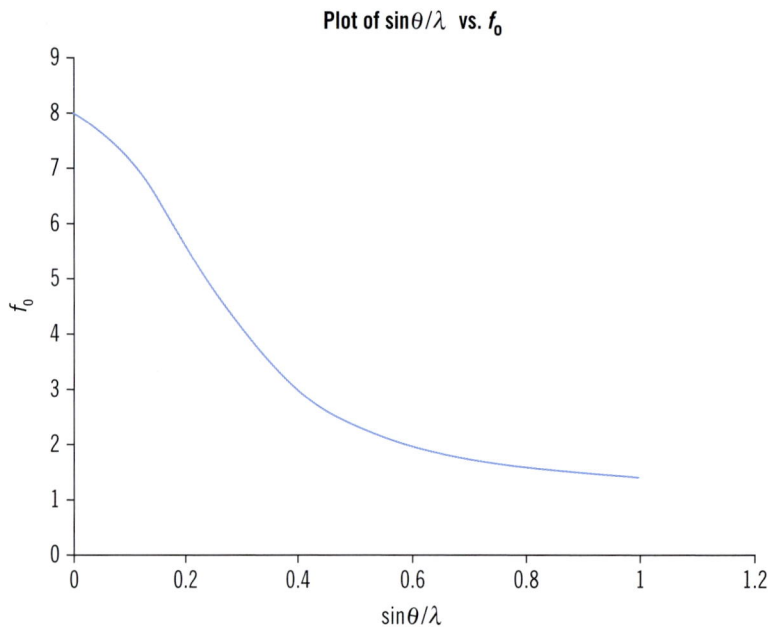

FIGURE 5.3 The atomic scattering factor of oxygen as a function of the Bragg angle

$$F_{hkl}^{calc} = \sum_{1}^{N} f_n \cos 2\pi(hx_n + ky_n + lz_n) + i\sum_{1}^{N} f_n \sin 2\pi(hx_n + ky_n + lz_n),$$

The calculation of the structure factor, F_{hkl}^{calc} can then be divided into two sections: the right-hand side of the equation splits easily into two halves, the summation with the cosine function and the summation with the sine function, which consider the atomic scattering factors of the atoms within the unit cell; with the sine function having an imaginary component $i = \sqrt{-1}$.

Based on scattering within a unit cell:

$$F_{hkl}^{calc} = \sum_{1}^{N} f_n e^{i\phi_n} = \sum_{1}^{N} f_n e^{2\pi i(hx_n + ky_n + lz_n)}$$

$$e^{ix} = \cos x + i \sin x,$$

so, substituting $\cos x + i \sin x$ for e^{ix},

$$F_{hkl}^{calc} = \sum_{1}^{N} f_n \cos 2\pi(hx_n + ky_n + lz_n) + i\sum_{1}^{N} f_n \sin 2\pi(hx_n + ky_n + lz_n).$$

In a diffraction pattern, an absence occurs when the intensity of a reflection from a specific set of Miller planes is equal to zero, $I = 0$. As the intensity, I, is equal to the square of the structure factor, F, then by definition, in an absence, F^{calc}, is also equal to zero.

Considering a centrosymmetric unit cell (a unit cell that possesses a centre of symmetry), the sine part of the equation disappears. So, F_{hkl}^{calc} for a centrosymmetric unit cell becomes just

$$F_{hkl}^{calc} = \sum_{1}^{N} f_n \cos 2\pi (hx_n + ky_n + lz_n) \qquad (5.2)$$

This is because, in a centrosymmetric unit cell, atoms will be located at x, y, z and at $-x, -y, -z$. Substituting the values into Eqn 5.1, we find that $\sin(-x) = -\sin(x)$. We also find that $\cos 2\pi$ of an odd number is always equal to -1 while $\cos 2\pi$ of an even number is always equal to $+1$.

$\cos 2\pi(\text{odd number}) = -1$
$\cos 2\pi(\text{even number}) = +1$

Consider the following example of how structure factors are calculated in a simple centrosymmetric unit cell.

▷ **WORKED EXAMPLE: CALCULATING STRUCTURE FACTORS**

Scenario

A unit cell contains two atoms of carbon with fractional coordinates (0.10, 0.10, 0.10), (0.90, 0.90, 0.90) and two atoms of oxygen with coordinates (0.20, 0.85, 0.30), (0.80, 0.15, 0.70).

The parameters of the unit cell are $a = b = 3$ Å, $c = 5$ Å, and $\alpha = \beta = \gamma = 90°$. The scattering factors for the two types of atom are given below as functions of $(\sin\theta)/\lambda$.

$(\sin\theta)/\lambda$	f_C	f_O
0	6	8
0.1	5.13	7.25
0.2	3.58	5.63
0.3	2.5	4.09
0.4	1.95	3.01
0.5	1.69	2.34
0.6	1.54	1.94

With the information provided calculate the structure factor, F_{112}, and subsequently determine the equivalent intensity value, I.

Strategy for solution

With these data, we can calculate the F_{hkl} values for all reflections, such that $h^2 + k^2 + l^2 < 8$ and all Miller indices are positive. As the structure is centrosymmetric (each atom pair is related by an inversion centre at ($\frac{1}{2}$, $\frac{1}{2}$, $\frac{1}{2}$)) all F are real (i.e., the sine function of the equation disappears) leaving:

$$F_{hkl}^{calc} = \sum_{1}^{N} f_n \cos 2\pi(hx_n + ky_n + lz_n).$$

1. First, we need to determine each atomic scattering factor (f_x) at the relevant $(\sin\theta)/\lambda$ for the given Miller index, in this case (1, 1, 2).

2. Next, determine $\cos 2\pi(hx_n + ky_n + lz_n)$ for each atom pair.

3. Finally, calculate F_{112} by substituting the values into the equation.

Calculate $(\sin\theta)/\lambda$ for Miller index (1, 1, 2) and subsequently determine the atomic scattering factors

From Bragg's law;

$$2d\sin\theta = n\lambda$$

$$\frac{2\sin\theta}{\lambda} = \frac{n}{d}.$$

Hence,

$$\frac{4\sin^2\theta}{\lambda^2} = \frac{h^2}{a^2} + \frac{k^2}{b^2} + \frac{l^2}{c^2}.$$

For F_{112},

$$\frac{4\sin^2\theta}{\lambda^2} = \frac{1}{3} + \frac{1}{3} + \frac{1}{25}$$
$$= 0.3822.$$

Therefore,

$(\sin\theta)/\lambda = \sqrt{(0.3822)}/2 = 0.6182/2 = 0.31.$

From graphs of f_C and f_O versus $(\sin\theta)/\lambda$,

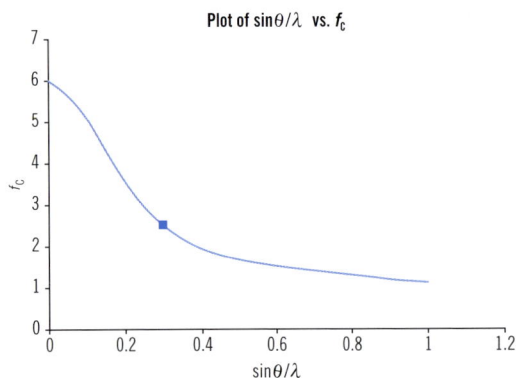

Plot of $\sin\theta/\lambda$ vs. f_c

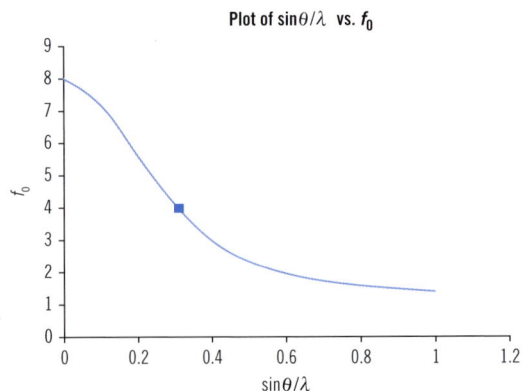

Plot of $\sin\theta/\lambda$ vs. f_0

we find that when $(\sin\theta)/\lambda = 0.31$, $f_C = 2.5$ and $f_O = 3.97$.

Determine cos2π(hx + ky + lz) for each atom pair

For the first atom pair:

$$hx_C + ky_C + lz_C = 0.40; \quad hx_O + ky_O + lz_O = 1.65;$$

$$\cos2\pi(hx_C + ky_C + lz_C) = -0.81; \quad \cos2\pi(hx_O + ky_O + lz_O) = -0.59.$$

While for the second atom pair:

$$hx_C + ky_C + lz_C = 3.6; \quad hx_O + ky_O + lz_O = 2.35;$$

$$\cos2\pi(hx_C + ky_C + lz_C) = -0.81; \quad \cos2\pi(hx_0 + ky_0 + lz_0) = -0.59.$$

Determining the structure factor (F_{112}) associated with Miller index (1, 1, 2)

$$F_{112} = \sum_{1}^{N} f_n \cos2\pi(hx_n + ky_n + lz_n)$$

$$= [(2.5\times(-0.81)) + (2.5\times(-0.81))] + [(3.97\times(-0.59)) + (3.97\times(-0.59))].$$

$$= -8.70 \text{ (correct to two decimal places)}$$

The (almost) completed table follows (to two decimal places):

h	k	l	F	$I = F^2$
1	0	0	10.36	107.32
0	0	1	3.90	15.17
1	1	0	11.46	131.22
0	1	1	9.10	82.87
1	0	1	-9.21	84.85
1	1	1	-7.21	51.96
2	0	0	-4.63	21.45
0	2	0	-0.85	0.72
0	0	2	-7.01	49.08
2	1	0	-1.31	1.70

h	k	l	F	$I = F^2$
1	0	2	1.09	1.19
1	2	0	4.14	17.17
0	1	2	-10.56	111.51
0	2	1	5.59	31.25
2	1	1	-9.19	84.51
1	**1**	**2**	**-8.70**	**75.77**
1	2	1		
2	2	0		
2	0	2		
0	2	2		

SELF-TEST QUESTIONS

1. Describe the term *absence,* in terms of intensity, *I,* and the diffraction map.

2. Briefly describe the process that leads to the production of an *.hkl* file.

3. Explain how the atomic scattering factors are influenced by Bragg's angle.

4. Why is it sometimes difficult to locate the light atoms in an electron-density map when heavy elements are present in the crystal structure?

5. Calculate the remaining four *F* values, and subsequently work out the calculated intensities (prior to applied corrections) for each set of Miller indices. (Tip: your calculator has to be in the radian mode.)

5.3 SPACE GROUPS AND ABSENCES

Section learning outcomes

To be able to:

- Recognize and identify general absences and systematic absences;
- Understand and explain how general absences are derived;
- Determine space groups from general and systematic absences;
- Identify absences in a dataset.

An essential part to solving a crystal structure relies on determining the lattice type and symmetry elements (or space groups) within the crystal lattice. This information is contained within the diffraction pattern of each crystal lattice, where both reflections and absences contribute. As we saw from the earlier sections, the structure factors contain information relating to the Miller planes and the location (fractional coordinates) of each atom type within the unit cell. By examining the diffraction pattern's reflections and the absences within the unit cell, one is able to determine the location of all the atoms and molecules within it.

Absences in diffraction patterns (when $I = 0$), can be divided into two categories; *general absences*, pertaining to lattice types, and *systematic absences*, relating to translational symmetry elements.

5.3.1 General absences

On a diffraction pattern, destructive interference of X-rays can occur, resulting in absences for *non-primitive* lattice types (for example, I, F, or C). The rules defining these absences affect all the reflections in the diffraction pattern (for all Miller index values of *hkl*). These absences are known as general absences.

Apart from the primitive lattice type (P) all other lattice types can be determined by general absences that occur within the diffraction pattern. These absences are defined by rules relating to certain sets of Miller indices within the diffraction data, as given in Table 5.2.

For the body-centred (I) lattice type, absences occur when the sum of the Miller indices $(h + k + l)$ is an odd value, while for the face-centred (F) lattice types, absences occur when all indices (*h*, *k*, and *l* values) are either all even or all odd.

For general absences relating to the specific face-centred lattices, absences occur when the sum of the two non-corresponding Miller indices are an odd value. For example, for

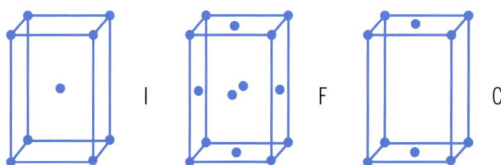

FIGURE 5.4 Some non-primitive lattice types

TABLE 5.2 Rules governing general absences

Lattice type	Conditions for general absences
P	None
A	$k + l = 2n + 1$ (i.e., the sum of k and l is odd)
B	$h + l = 2n + 1$ (i.e., the sum of h and l is odd)
C	$h + k = 2n + 1$ (i.e., the sum of h and k is odd)
F	Reflections must have either all even or all odd indices to be observed Mixed odd and even indices are not allowed
I	$h + k + l = 2n + 1$ (i.e., the sum of the indices is odd)

the A-face centred lattice, absences occur when the sum of k and l is an odd value. For the B-face centred lattice, the absences occur when the sum of h and l is an odd value, while for the C-face centred lattice, absences occur when $h + k$ is an odd value.

By calculating the structure factors and considering the coordinates of the unit cell lattice points, we are able to show how the rules for general absences are derived.

▷ **GENERAL ABSENCE SELECTION RULE FOR AN A-TYPE BRAVAIS LATTICE – AN EXAMPLE**

In the example of an A-type Bravais lattice, assuming that each lattice point is equivalent to a single atom of type J, the coordinates of atoms at the eight corners of the unit cell are $(0, 0, 0)$, $(1, 0, 0)$, $(0, 1, 0)$, $(0, 0, 1)$, $(1, 1, 0)$, $(0, 1, 1)$, $(1, 0, 1)$, and $(1, 1, 1)$. Each atom makes a 1/8 contribution to the unit cell.

As the lattice points (in this case atoms of type J) at the corners of the unit cell are all equivalent, any one of these can be selected to count as the total contribution from all the corners of the unit cell.

Let us select the atom at $(0, 0, 0)$.

The atoms at the centres of the two A faces of the unit cell have fractional coordinates $(0, \frac{1}{2}, \frac{1}{2})$ and $(1, \frac{1}{2}, \frac{1}{2})$. Each of these atoms makes a contribution of $\frac{1}{2}$ to the unit cell. As these atoms are also equivalent, either of these can be selected as the total contribution from the corners of the unit cell. Let us select the atom at $(0, \frac{1}{2}, \frac{1}{2})$.

Substituting into Eqn 5.2, we get an expression for the general structure factor for this Bravais lattice.

$$F_{hkl} = f_J \cos 2\pi(0) + f_J \cos 2\pi \left(\frac{1}{2}k + \frac{1}{2}l \right).$$

Simplifying the equation:

$$F_{hkl} = f_J + f_J \cos \pi(k + l).$$

Keeping in mind that when an absence occurs, the intensity, I, is equal to zero ($I = 0$). (Remember also that $I = F^2$)

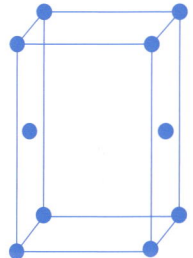

From the equation, it becomes clear that the structure factors will have a value of zero when $\cos\pi(k+l) = -1$. This will only occur when $k + l$ is an odd number. Thus, an A-type Bravais lattice will be identified when all reflections in a dataset for which $k + l$ is an odd number are absent, i.e., have no intensity. This is often written as $k + l = 2n + 1$.

A similar method can be used for proving other lattice types. It can also be extended to more complex cases in which the lattice point environment is a molecule or a group of molecules.

Examples of absences corresponding to A-type lattices are for hkl values of: (1, 0, 1), (1, 2, 1), (2, 3, 2), and (3, 2, 5).

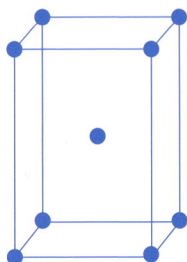

▷ GENERAL ABSENCE SELECTION RULE FOR AN I-TYPE BRAVAIS LATTICE – AN EXAMPLE

In the example of an I-type (body-centred) Bravais lattice, assuming that each lattice point is equivalent to a single atom of type J, the coordinates of atoms at the eight corners of the unit cell are (0, 0, 0), (1, 0, 0), (0, 1, 0), (0, 0, 1), (1, 1, 0), (0, 1, 1), (1, 0, 1), and (1, 1, 1). Each atom makes a 1/8 contribution to the unit cell.

As the lattice points (in this case atoms of type J) at the corners of the unit cell are all equivalent, any one of these can be selected to count as the total contribution from all the corners of the unit cell.

Let us select the atom at (0, 0, 0).

The atom in the centre of the unit cell has a fractional coordinate of (½, ½, ½). This atom makes a contribution of 1 to the unit cell.

Substituting into Eqn 5.2, we get an expression for the general structure factor for this Bravais lattice.

$$F_{hkl}^{calc} = \sum_{1}^{N} f_n \cos 2\pi \left(hx_n + ky_n + lz_n\right)$$

$$F_{hkl} = f_J \cos 2\pi (0) + f_J \cos 2\pi \left(\frac{1}{2}h + \frac{1}{2}k + \frac{1}{2}l\right).$$

Simplifying the equation:

$$F_{hkl} = f_J + f_J \cos\pi(h + k + l).$$

Similarly, the intensity, I, is equal to zero ($I = 0$) when an absence occurs. (Remember also that $I = F^2$.)

From the equation, it becomes clear that the structure factors will have a value of zero when $\cos\pi(h + k + l) = -1$. This will only occur when $h + k + l$ is an odd number.

Therefore, an I-type Bravais lattice will be identified when all reflections in a dataset for which $h + k + l$ is an odd number are absent, i.e., have no intensity. Similarly, this is often written as $h + k + l = 2n + 1$.

Examples of absences corresponding to I-type lattices are for hkl values of: (1, 1, 1), (1, 3, 1), (2, 3, 2), and (3, 2, 6).

5.3.2 Systematic absences

Like general absences, systematic absences in diffraction data arise when destructive interference of X-rays occurs in relation to specific translational symmetry elements (screw axes and glide planes). The absences also take place according to rules relating to specific sets of Miller indices within the diffraction data.

Systematic absences can be found in diffraction data only for translational symmetry elements, screw axes, and glide planes. Systematic absences do not occur for non-translational symmetry elements.

Rules for systematic absences can be derived and proved using a similar method to the proof for general absences; however, this is beyond the scope of this book.

Let us now consider systematic absences that are characteristic of screw axes and glide planes.

Screw axes

The rules determining systematic absences associated with screw axes (see Table 5.3) are based on the axis to which the screw is parallel. For example, for screw axes along a, the condition for the systematic absence is always a derivative of the $(h, 0, 0)$ Miller index, where an h value is odd and absent.

A 2_1 screw axis along a would have multiples of $(2n + 1)$ values of h absent (see Table 5.4). This similarly occurs for the 4_1 axis. However, reflections corresponding to $(2n + 1)$ will also be absent as this is a subset of $(4n + 1)$. This similarly occurs for the 3_n and 6_n screw axes.

Similarly, screw axes along b are derivatives of $(0, k, 0)$; with odd values of k being absent, and screw axes along c are derivatives of $(0, 0, l)$ where values of l that are odd are absent.

Glide planes

In determining the systematic absences rules for glide planes, we consider the following: The equivalent Miller index of the axes to which the glide is perpendicular is always 0. It is the axis along which the glide plane is parallel that defines the conditions for the absence. The rules are given in Table 5.5.

For example, a glide plane perpendicular to the a-axis is given by $(0, k, l)$; if the glide plane is along the b-axis, then the values for the absent reflections are defined by the k

TABLE 5.3 Systematic absences for screw axes

Symmetry element		Conditions for systematic absence	
Screw axis along a	$2_1, 4_2$	$(h, 0, 0)$	$h = 2n + 1$
	$4_1, 4_3$	$(h, 0, 0)$	$h = 4n + 1$
Screw axis along b	$2_1, 4_2$	$(0, k, 0)$	$k = 2n + 1$
	$4_1, 4_3$	$(0, k, 0)$	$k = 4n + 1$
Screw axis along c	$2_1, 4_2, 6_3$	$(0, 0, l)$	$l = 2n + 1$
	$3_1, 3_2, 6_2, 6_4$	$(0, 0, l)$	$l = 3n + 1$
	$4_1, 4_3$	$(0, 0, l)$	$l = 4n + 1$
	$6_1, 6_5$	$(0, 0, l)$	$l = 6n + 1$

TABLE 5.4 A sample dataset – systematic absences caused by a 2_1 screw axis along c

h	k	l	I_{hkl}	σ_{hkl}	h	k	l	I_{hkl}	σ_{hkl}
0	0	1	6.4	0.2	0	0	12	82.5	0.8
0	0	2	135.0	0.5	0	0	13	0.4	0.2
0	0	3	0.6	0.1	0	0	14	98.1	1.3
0	0	4	254.8	1.2	0	0	15	0.2	0.1
0	0	5	0.1	0.6	0	0	16	101.1	1.2
0	0	6	194.2	1.4	0	0	17	0.6	0.7
0	0	7	0.7	0.5	0	0	18	0.9	1.6
0	0	8	376.1	2.5	0	0	19	0.5	0.2
0	0	9	0.3	0.7	0	0	20	54.2	0.7
0	0	10	253.4	1.2	0	0	21	0.7	0.6
0	0	11	0.6	0.3	0	0	22	98.8	1.2

TABLE 5.5 Systematic absences for glide planes

Symmetry element			Conditions for systematic absence	
Glide plane II (1, 0, 0)	b glide perpendicular a		$(0, k, l)$	$k = 2n + 1$
	c glide perpendicular a		$(0, k, l)$	$l = 2n + 1$
Glide plane II (0, 1, 0)	a glide perpendicular b		$(h, 0, l)$	$h = 2n + 1$
	c glide perpendicular b		$(h, 0, l)$	$l = 2n + 1$
Glide plane II (0, 0, 1)	a glide perpendicular c		$(h, k, 0)$	$h = 2n + 1$
	b glide perpendicular c		$(h, k, 0)$	$k = 2n + 1$

TABLE 5.6 A sample dataset – systematic absences caused by an a glide perpendicular to b

h	k	l	I_{hkl}	σ_{hkl}	h	k	l	I_{hkl}	σ_{hkl}
6	0	1	253.6	0.1	−4	0	1	324.7	1.1
5	0	1	0.1	0.6	−5	0	1	2.5	0.5
4	0	1	132.8	0.4	−6	0	2	523.8	0.2
3	0	1	0.7	0.3	−7	0	2	0.8	0.1
2	0	1	312.2	1.2	−6	0	2	34.2	0.7
1	0	1	0.2	0.7	−5	0	2	0.4	0.2
0	0	1	156.1	0.9	−4	0	2	257.2	1.5
−1	0	1	−0.1	0.2	−3	0	2	1.2	0.1
−2	0	1	65.1	2.1	−2	0	2	136.8	1.5
−3	0	1	1.6	0.5	−1	0	2	0.3	0.6
					0	0	2	112.8	2.1
					1	0	2	0.1	0.5
					2	0	2	86.1	1.1

values being odd and absent. For a *b* glide perpendicular to *a*, systematic absences are defined for $(0, k, l)$, with odd values of *k* being absent $(k = 2n + 1)$.

If we examine the sample dataset in Table 5.6, we find that the dataset represents values of $(h, 0, l)$, which suggests a glide plane perpendicular to the *b*-axis. If we examine the table for absences, we find that absences occur for all values of *h* that are odd, which suggests an *a* glide. We can then conclude that the table represents absences, indicating that within this particular crystal structure there is an *a* glide that lies perpendicular to the *b*-axis.

SELF-TEST QUESTIONS

1. Based on the example for the A-type Bravais lattice in Section 5.3.1, prove that the general absences rule for the C-type Bravais lattice is $(h + k)$ absent for all odd values.

2. Briefly list the general absences you would expect to find if the crystal lattice was a:

   (a) B-type Bravais lattice;

   (b) F-type Bravais lattice;

   (c) I-type Bravais lattice.

3. Determine the space group for a crystal lattice that exhibits the following absences:

   (a) No general absences;

   (b) Systematic absences corresponding to:

   $(0, k, l)$, where values of k odd are absent;

   $(h, 0, l)$, where values of l odd are absent;

   $(h, k, 0)$, where values of h odd are absent.

4. What absences would you expect to find for a crystal structure with the orthorhombic space group $P\,2_1\,2_1\,2_1$?

⊡ CHAPTER SUMMARY

1. An absence occurs when diffraction spots or reflections are 'missing' from diffraction data.

2. Absences provide information about Bravais lattice types (general absences) and non-translational symmentry – screw axes and glide planes (systematic absences)

3. The intensity, I, is a function of the square of the structure factor.

4. The structure factor is derived from atomic scattering factors.

5. Atomic scattering factors are dependent on the electron density of the atom and are influenced by the Bragg angle of the incident X-ray.

6. The rules governing general absences are:

Lattice type	Conditions for general absences
P	None
A	$k + l = 2n + 1$ (i.e., sum of $k + l$ odd)
B	$h + l = 2n + 1$ (i.e., sum of $h + l$ odd)
C	$h + k = 2n + 1$ (i.e., sum of $h + k$ odd)
F	Reflections must have either all even or all odd indices to be observed. Mixed odd and even indices are not allowed.
I	$h + k + l = 2n + 1$ (i.e., sum of indices odd)

7. The rules governing systematic absences are:

Symmetry element			Conditions for systematic absence	
Screw axis along a	$2_1, 4_2$		$(h, 0, 0)$	$h = 2n + 1$
	$4_1, 4_3$		$(h, 0, 0)$	$h = 4n + 1$
Screw axis along b	$2_1, 4_2$		$(0, k, 0)$	$k = 2n + 1$
	$4_1, 4_3$		$(0, k, 0)$	$k = 4n + 1$
Screw axis along c	$2_1, 4_2, 6_3$		$(0, 0, l)$	$l = 2n + 1$
	$3_1, 3_2, 6_2, 6_4$		$(0, 0, l)$	$l = 3n + 1$
	$4_1, 4_3$		$(0, 0, l)$	$l = 4n + 1$
	$6_1, 6_5$		$(0, 0, l)$	$l = 6n + 1$

Symmetry element		Conditions for systematic absence	
Glide plane ‖ (1, 0, 0)	b glide perpendicular a	$(0, k, l)$	$k = 2n + 1$
	c glide perpendicular a	$(0, k, l)$	$l = 2n + 1$
Glide plane ‖ (0, 1, 0)	a glide perpendicular b	$(h, 0, l)$	$h = 2n + 1$
	c glide perpendicular b	$(h, 0, l)$	$l = 2n + 1$
Glide plane ‖ (0, 0, 1)	a glide perpendicular c	$(h, k, 0)$	$h = 2n + 1$
	b glide perpendicular c	$(h, k, 0)$	$k = 2n + 1$

FURTHER READING

Giacovazzo, C., Monaco, H. L., Artioli, G., *et al.* (eds.) (2002). *Fundamentals of Crystallography,* 2nd edn. IUCr Texts on Crystallography. Oxford University Press, New York.

Glusker, J. P. and Trueblood, K. N. (1985). *Crystal Structure Analysis – A Primer.* Oxford University Press, New York.

Stout, G. H. and Jensen, L. H. (1989). *X-Ray Structure Determination: A Practical Guide.* John Wiley & Sons, Ltd, New York.

STRUCTURE SOLUTION

6

By the end of this chapter you should be able to:

- **Outline the phase problem;**
- **Understand the concept of Fourier transformation and how it relates real space parameters to reciprocal space parameters;**
- **Describe the relevant methods used to overcome the phase problem;**
- **Identify the best structure solution method for a particular crystal.**

6.1 INTRODUCTION

Structure solution, also known as *phasing*, is the process in which a molecular model can be calculated and created so that it may represent the contents of the crystal lattice under investigation. This process of obtaining a model for a crystal structure involves solving the phase problem.

To understand the phase problem better, we need to recall from Chapter 5 how the intensity of each diffraction spot is related to the structure factor and, consequently, the atomic scattering factors of the atoms in the crystal.

Reminder:

$$I_{hkl}^{corr} \propto \left(F_{hkl}^{obs} \right)^2,$$

$$F_{hkl}^{calc} = \sum_1^N f_n \cos 2\pi \left(hx_n + ky_n + lz_n \right) + i \sum_1^N f_n \sin 2\pi \left(hx_n + ky_n + lz_n \right)$$

The value of the corrected intensity I^{corr} (of a diffraction spot) as a function of the square of the structure factor $(F^{obs})^2$ is always a positive value.

On a diffraction pattern, each diffraction spot, each value of intensity, and, subsequently, each structure factor, has associated with it the amplitude, the direction, and the *phase* angle of the incident X-ray.

While, we can identify both the amplitude (based on the intensity of the X-rays) and the direction (by establishing the set of Miller indices for each reflection); we can only properly find a crystal model once we are able to establish the X-ray *phase* angle of each reflection. To do this, we need to be able to determine the sign (whether positive or negative) for each structure factor. This is also known as the *phase problem*.

For centrosymmetric structures, that is, structures with a centre of symmetry at the origin, each reflection will have a phase angle of either 0° or 180° (+ or −). However, for a non-centrosymmetric structure, the phase angle for each reflection may take a value anywhere between the limits of 0° and 180°.

6.2 THE PHASE PROBLEM (FOURIER TRANSFORM)

> **Section learning outcomes**
>
> To be able to:
>
> • Describe the phase problem;
> • Explain how Fourier transform is used to represent the phase problem.

6.2.1 The phase problem: visually explained

If we refer to Fig. 6.1 and observe the interaction between the X-ray source and a single crystal, we see that an X-ray source produces a continuous source of X-rays. These X-rays, in the form of X-ray waves, reach the single crystal in phases and various combinations of phases. However as the crystal diffracts the X-rays, a phase shift may occur and when the diffraction pattern is subsequently recorded, it contains information on the amplitude and the direction of the incident X-ray waves but the phase information is lost.

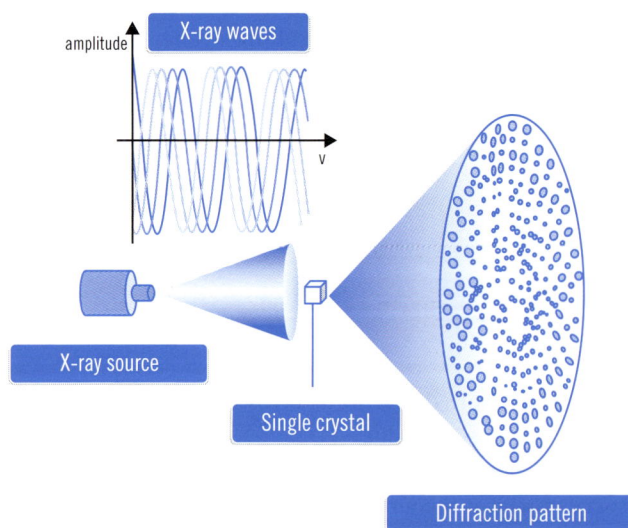

FIGURE 6.1 A continuous source of X-rays

To solve a crystal structure and to identify the positions of each atom within the crystal, it is necessary to identify the phases of the X-rays, that is, the signs and the values of the phases for the structure factors. The phases are determined by the angle at which the incident X-ray reaches the crystal (based on Bragg's law). This is also commonly termed the *phase angle*.

Mathematically, we can relate the electron density within a crystal lattice, and subsequently the phase problem, to the Fourier series. As we have seen in Chapter 1, the crystal lattice consists of a large number of repeating unit cells; this is considered to be a periodic function. The Fourier series can be used to represent mathematically any periodic function that uses the trigonometric functions of sine and cosine.

Based on the Fourier series, the electron density at any point (x, y, z) within a crystal lattice can be represented by the following Fourier equation:

$$\rho_{xyz} = \frac{1}{V}\sum_h\sum_k\sum_l |F_{hkl}| \cos 2\pi(hx + ky + lz - \alpha_{hkl}).$$

The electron density, ρ_{xyz}, at location (x, y, z) in a unit cell with volume, V, is expressed in terms of electrons per cubic angstroms (e Å^{-3}). The equation comprises the summation of the amplitudes of the structure factors, F_{hkl}, and the relative (as yet unknown) phase angle α_{hkl} of each Bragg reflection.

Based on this equation, the Fourier transform gives us a way of representing the experimental Bragg reflections in terms of an electron density map, where the electron density is a Fourier transform of the structure factors, amplitudes, and phases at a point (x, y, z) within the crystal lattice. Each structure factor, F_{hkl}, contains information about the positions of all atoms from all symmetry-related objects within the unit cell. Once these atomic positions can be determined, the phases of all the structure factors can be calculated.

The Fourier series is named after **Jean Baptiste Joseph Fourier** (1768–1830). Fourier was a French mathematician and physicist, and is best known for initiating the investigation of Fourier series and its application to the problems of heat flow.

The Fourier series consists of a Fourier analysis (forward Fourier transform) and a Fourier synthesis (reverse Fourier transform). The Fourier analysis relates the physical crystal lattice to an electron density map, while the Fourier synthesis resolves the electron density map to a crystal structure.

TABLE 6.1 Real space and reciprocal space related by the Fourier series

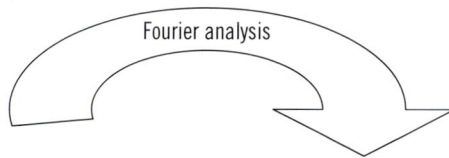

Fourier analysis

Real space	Reciprocal space
Crystal structure	Diffraction pattern
Electron density (atomic parameters)	Structure factors, Amplitudes and phases (X-rays)
Crystal lattice; unit cell	Reciprocal lattice; cell
Coordinates (x, y, z)	Coordinates (h, k, l)

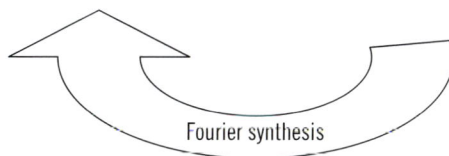

Fourier synthesis

To the right of the Fourier equation, we can determine the values for all except the phase angles (α_{hkl}). Once the phases are assigned through the structure solution process, the electron density at any point (x, y, z) within the crystal lattice can be calculated. A molecular model of the contents of the crystal lattice can then be derived.

Table 6.1 outlines how real space and reciprocal space are related by the Fourier series; the Fourier synthesis, and the Fourier analysis.

$$\rho_{xyz} = \frac{1}{V}\sum_{h}\sum_{k}\sum_{l}\left|F_{hkl}\right|\cos 2\pi(hx+ky+lz-\alpha_{hkl}).$$

Now that we have considered the phase problem, in the following sections we will examine how we are able to approach the problem and seek relevant solutions.

The process of solving the phase problem is also commonly referred to as the structure solution process in crystallography. While many of the fundamentals of crystallography remain congruous between small-molecule and macromolecular crystallography, here we find however, that the structure solutions diverge between small-molecule crystallography and macromolecular crystallography and we will examine each in turn.

> ### SELF-TEST QUESTIONS
>
> 1. What information is required to construct a model of a crystal structure?
>
> 2. Briefly outline the phase problem and explain how the Fourier transform can be used to represent it.
>
> 3. How is real space related to reciprocal space by the Fourier transform?

Prior to attempting to solve a structure it is imperative that the crystal structure is assigned the correct space group. If the space group is not correctly assigned, it is often extremely difficult and perhaps even impossible to solve the crystal structure. It is also often very helpful to have an idea of what the molecular contents of the crystal may be, so that the features within the experimental data may be more easily recognized. The initial model obtained from the structure solution process is found only for the asymmetric unit of the crystal lattice and contains only non-hydrogen atoms.

6.3 SMALL MOLECULES: STRUCTURE SOLUTION

> ### Section learning outcomes
>
> To be able to:
>
> - Identify the structure solution methods for small molecules;
>
> - Sketch the processes involved in direct methods and Patterson synthesis.

Methods used to overcome the phase problem, or structure solution methods in small-molecule crystallography, are largely based on two approaches, direct methods and Patterson synthesis. Both methods are used for different types of molecule and in the overall crystallographic process, each is used only once. Once a solution is achieved, an electron density map is drawn and this provides a good representation of where and how the electron density is distributed.

6.3.1 Direct methods

Direct methods are so named because they require that the crystallographer (the computer in current times) guess or estimate *directly* the possible phase angles (of each Bragg reflection) for each value of the structure factors, F_{hkl}. Historically, the direct methods were used mainly when a crystal structure was known to be an organic molecule and contained no heavy elements within the crystal lattice. However, the development of the theory of direct methods over the last three decades, and the availability of computer software, has enabled the use of direct methods to solve more and more complex structures, including metal-containing complexes and some proteins.

Two important conditions are imposed on the direct methods and are always considered in the course of applying this method of structure solution. They are:

(a) That the value of electron density (ρ) must never be negative; it may be positive or zero but never negative; and

(b) That the electron density maps have to have high values (sharp peaks) at and near atomic positions and must have nearly zero values everywhere else.

In considering these two conditions we can then hope to calculate structure factors, F_{hkl}, with the correct value of the phase angles, α_{hkl}.

The process of structure solution using the direct methods initially uses *normalized structure factors*; these are obtained when the measured structure factors are modified to obtain the maximum information on atomic position, assuming that there is no temperature contribution to the structure factor. Only the normalized structure factors with the highest values are used, and these can be represented by E_{hkl}.

E_{hkl} can be represented by:

$$E_{hkl} \propto \frac{F_{hkl}}{\sqrt{\sum_i f_i}},$$

where F_{hkl} is modified to take into account the fall-off in individual scattering factors, f, with the increasing Bragg angle 2θ.
Reminder:

$$I_{hkl}^{corr} \propto \left(F_{hkl}^{obs}\right)^2.$$

Once these normalized structure factors are found, phases are assigned to some of the strongest reflections and these are used as a basis of obtaining the initial locations of the origin $(0, 0, 0)$ of a unit cell within the crystal lattice. The location of the origin plays an important role in determining the relative phases of the Bragg reflections once the initial phases are assigned.

Once the origin is determined, the reflections are examined in groups or sets of three (known as a triplet set), and their phase angles compared. The triplet sets of reflections, $h_1 k_1 l_1$, $h_2 k_2 l_2$, and $h_3 k_3 l_3$, are such that $h_1 + h_2 + h_3 = k_1 + k_2 + k_3 = l_1 + l_2 + l_3 = 0$. An example of a triplet set would be the Bragg reflections $33\bar{1}$ $1\bar{6}0$ $\bar{4}31$.

The probabilities for each triplet set are then calculated such that each product is positive and each E_{hkl} value is positive.

This leads to the mapping of a 'most probable' phase set that is used to calculate an E_{hkl} map, an electron density map based on normalized structure factors.

The map is then examined for recognizable features or familiar molecular fragments, such as phenyl rings or alkyl side chains. For example, in Fig. 6.2, we can clearly identify the hexagonal shape of a phenyl ring. This can then be used as an initial trial structure, where the positions of the atoms are put into the Fourier equation and the signs of the structure factors calculated.

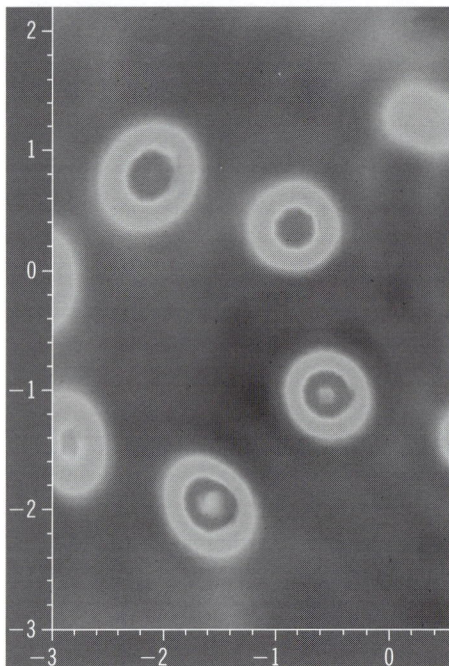

FIGURE 6.2 An example of an electron density map

Once the structure is solved in this way, the structure is then carried through to the refinement process. This is explained further in Chapter 7.

6.3.2 Patterson methods

Arthur Lindo Patterson (1902–1966), while at Massachusetts Institute of Technology in the USA in 1934, developed a method of solving crystal structures, which involves the summing of the Fourier series in two and three dimensions.

The Patterson synthesis or Patterson function is another approach to solving the phase problem, which essentially consists of using a trial structure to calculate the relative phase angles. This is very useful in determining initially the locations of the heavy atoms in a molecule and is often employed in inorganic and organometallic crystal structures. The Patterson function is also often used in macromolecular structure solutions, particularly when the crystals contain protein derivatives with a heavy metal.

The Patterson function is a map that draws all the possible relationships or vectors (considering both distance and direction) between all of the atoms in the crystal structure.

$$\rho_{xyz} = \frac{1}{V} \sum_h \sum_k \sum_l \left| F_{hkl} \right|^2 \cos 2\pi \, (hx + ky + lz).$$

The Patterson function, shown above, is also a Fourier series, however, as it takes the squared values of the structure factors; the need for the phase-angle information from the Fourier equation is removed. As the values of the structure factors can be found directly from the X-ray diffraction, the map of the Patterson function can be calculated with certainty.

The peaks in a Patterson map represent the vector between two atoms. If we refer to Fig. 6.3(a), for example, a peak located at (a, b, c) on the Patterson map means that there are two atoms separated by that vector, one at (x, y, z) and another at $(x + a, y + b, z + c)$, shown in Fig. 6.3(b).

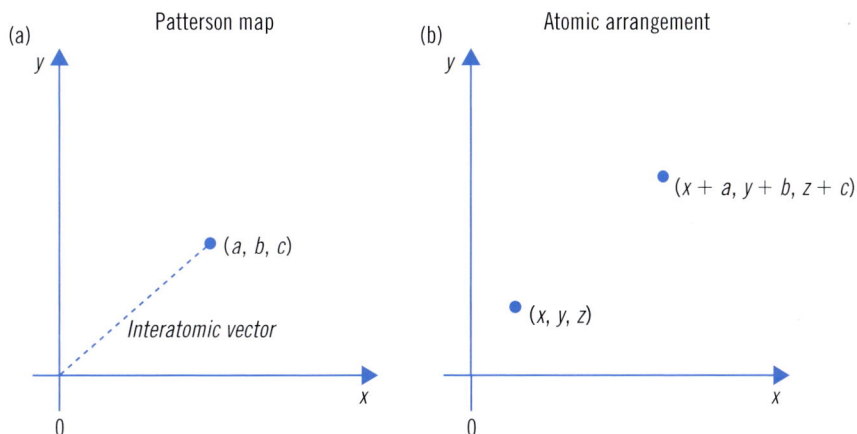

FIGURE 6.3 (a) The interatomic vector from the origin to peak (a, b, c) on a Patterson map; (b) the same interatomic vector between atoms at (x, y, z) and $(x + a, y + b, z + c)$

The highest peak in any Patterson map is at the origin as this is the vector between every atom and itself. However this peak is now often removed by the Patterson algorithms used.

The peak heights in a Patterson map are proportional to the product of the atomic numbers of the two atoms of the vector that is drawn. This means that the highest peak in the Patterson map would be drawn between the heaviest atoms in the crystal structure. Following this, the heaviest atom can then be located confidently, and the phase angles approximated within the Fourier equation. Subsequently all other atoms can then be located in turn and the structure refined.

SELF-TEST QUESTIONS

1. A chemist has synthesized a new organometallic molecular complex containing mercury, Hg and has managed to crystallize both the complex and its ligand separately. He would now like to determine the structures of both these crystals.

   (a) What methods of structure solution should the crystallographer use for each type of crystal? Explain why.

   (b) Briefly outline how each technique works.

2. Are there similarities between the two methods of small molecule structure solution? If so, what are they?

3. Outline the differences between the direct method and Patterson synthesis.

6.4 MACROMOLECULES: STRUCTURE SOLUTION

Like the structure solution process for small molecules, the main aim for structure solution in macromolecular crystallography is to solve the phase problem. In macromolecular crystallography, this process is commonly referred to as *phasing*.

Phasing for macromolecules takes a different approach from that for small molecules, where the identification of individual atoms take priority; for macromolecules, the general approach is the recognition of secondary, tertiary, and even quaternary structures (see Section 1.1.3). The techniques for phasing in macromolecular crystallography are direct methods, isomorphous replacement (IR), and multi-wavelength anomalous diffraction (MAD).

Section learning outcomes

To be able to:

- Identify the structure solution methods used in macromolecular crystallography;
- Outline the principles upon which the methods are based.

6.4.1 Direct methods

Structure solution by direct methods in macromolecular crystallography is similar to the process described in Section 6.3.1 for small-molecule crystallography. It is most commonly and most successfully used for the smaller protein molecules.

Another approach to direct methods, known as shake-and-bake, involves trying out a large number of random arrangements of atoms, simulating each of the diffraction patterns, and subsequently comparing the simulated pattern with that of the actual recorded pattern. The larger the structure, the higher the data resolution (better data) required (see Section 9.2.3). While this method can be successful, it relies significantly on computation power and its success is related directly to the computing power available.

6.4.2 Isomorphous replacement (IR)

The word *isomorphous* is derived from Greek; combining *isos* – meaning equal or the same – and *morphe* – meaning shape. Isomorphous refers to molecules that have similar atomic arrangements.

Isomorphous replacement is one of the most commonly used methods of structure solution in macromolecular crystallography. It is also sometimes used in the solution of isomorphous structures of small molecules.

The technique of IR uses the phases of a known solved protein structure as an initial model or as phase estimates for the new similar protein, where the phases of the known protein are used in the new set of experimental data as a phase model for the new protein. The amplitudes for the structure factors, F_{hkl} are taken from the experimental intensities of the new protein, while the phases are from the model. Further phase improvement is then carried through to the refinement process, as detailed in Chapter 7.

Isomorphous replacement is most favourably used in cases when (a) the new protein contains a derivative of the model or (b) when fragments of the model are replaced by other atoms. These are explained further in the two examples below.

▷ **EXAMPLE 1 – EXPERIMENTAL PROTEIN = KNOWN PROTEIN + LIGAND**

In the case when the new experimental protein contains a derivative of the known protein, it is then assumed that the protein folding is in a similar pattern. Isomorphous replacement is used to recognize and locate the protein, and then further phasing is carried out to locate the ligand.

▷ **EXAMPLE 2 – THE REPLACEMENT OF 'LIGHT' ATOMS BY 'HEAVY' ATOMS**

Protein crystals often contain between 30% and 90% of solvent molecules (or water) distributed in channels in the unit cell, it is then possible to replace the solvent or water in these channels with salts or other reactants that contain heavy atoms, by immersing the crystal in the relevant solutions. As the solvent molecules in the channels are replaced, the heavy atoms often interact with the side chains on the surface of the protein.

Isomorphous replacement in this situation relies on the multiple experimental determinations of several such derivatives of the same crystal structure. It relies on the *relative* phase determination, using several derivatives of original protein. The idea is that in the ideal situation, the original protein retains its shape and folds, and only the contents of the channels are changed (often assuming that the heavy atom replacements occur at the same points within the original structure) and that the overall crystal structure and packing is not changed significantly.

A simple equation represents this idea:

$$F_{PH} = F_p + F_H$$

where F_{PH} is the structure factor of the isomorphous replacement with the heavy metal derivative, F_P is that of the original protein and F_H is the contribution of the heavy atom. While we cannot measure F_H, as there are usually only a small number of heavy atom replacements, we can usually calculate it.

Generally, if the assumptions that the unit cell is unchanged are true, $F_H = F_{PH} - F_P$. Based on the heavy atom derivatives, the Patterson method can be used to determine positions of the heavier elements and subsequently phases of the original protein may be determined.

Using the IR method, further consideration can also be given to cases in which the new protein consists of rotations or translations of the known protein, or when it is sometimes necessary to modify the model, with minor deletions or additions (of side chains). In many of these cases, the model of the known protein can still be adapted for use.

As with other structure solution techniques, once a model is obtained the structure is then carried through to the refinement process.

6.4.3 Multi-wavelength anomalous diffraction (MAD)

Multi-wavelength anomalous diffraction (multiple anomalous dispersion), or MAD, as it is commonly known, is based on the principle of anomalous scattering.

All atoms and molecules, particularly heavy atoms, in the crystal absorb some of the X-rays and diffract the rest. We have so far assumed that real space is directly proportional to reciprocal space hence $hkl = -h -k -l$ (this is also known as Friedel's law); however, because of the crystal's ability to absorb X-rays, in many cases hkl and $-h -k -l$ are not equal.

This inequality in the incident and diffracted intensity is known as anomalous scattering or anomalous dispersion. In the MAD technique, the difference in the intensity in the Bijvoet pairs, (that is, the real and imaginary components, $|F_h(+)|^2$ and $|F_h(-)|^2$ can be exploited to overcome the phase problem. This is especially true when the crystal contains a heavy element that is able to produce sufficiently strong anomalous absorption.

The absorption capability for the elements is greatly reduced when the wavelengths are just below the characteristic emission wavelength of that particular element ($K\beta$) – this is known as the absorption edge and is described in terms of λ.

The wavelengths commonly used in X-ray experiments are not near the absorption edges of the most common elements (C, N, and O), therefore, these do not contribute to anomalous scattering. Heavy atoms, however, are in this absorption range and an X-ray diffraction experiment can be set up to optimize and take advantage of this unequal scattering.

If we remind ourselves that the structure factor is reliant on or related to, atomic scattering (see Section 5.2.1), we will also remember that the structure factor is dependent on the wavelength of the incident X-ray. We then find that the resulting diffraction pattern is greatly influenced by the differing wavelengths used in the MAD technique; the atomic differences, particularly those of the heavy atoms, can be enhanced.

A MAD experiment is usually run several times using a range of wavelengths on the same crystal, thereby obtaining several datasets. For example, a MAD experiment can be carried out at:

(a) Wavelength λ_1, 'the edge', where (real component) $F(+)$ is at its minimum;

(b) Wavelength λ_2, 'the white line', where (imaginary component) $F(-)$ is at its maximum and the Bijvoet differences between $F(+)$ and $F(-)$ are at their maximum;

(c) Wavelength λ_3, far from the edge, where both Fs are small.

This allows the identification of the heaviest atoms within the crystal structure, from a Patterson map or by direct methods. Once the phase angles have been determined for the heaviest atoms, the solution of the rest of the structure follows. This is then followed by the refinement process.

The use of synchrotron radiation (explained in further detail in Section 8.4.1) for MAD experiments is advantageous, as the wavelengths of the X-rays produced by the synchrotron can be tuned either towards or away from the absorption edge of a relevant element.

SELF-TEST QUESTIONS

1. Name the approaches a macromolecular crystallographer may use to identify the phases of a crystal.

2. In macromolecular crystallography, there are two direct methods approach to phasing: briefly compare the two.

3. What does IR stand for and what are its main principles?

4. Briefly outline anomalous scattering, which forms the basis of the MAD technique.

◉ CHAPTER SUMMARY

1. The structure solution process involves solving the phase problem.

2. While the information relating to the amplitude of the incident X-rays can be obtained

from the diffraction pattern, the phase information is lost – this is termed the phase problem.

3. A Fourier series can be used to represent the phase problem.

Small molecules – structure solution

4. Direct methods and the Patterson synthesis are used to solve small molecule structures.

5. Each method is used only once to obtain a structure solution.

6. Direct methods work best for organic molecules that do not contain heavy elements. They are based on 'guessing' the phases.

7. The two important conditions that must be fulfilled for direct methods are:

   (a) The electron density must never be negative;

   (b) In electron density maps, areas near atomic positions have high values while the electron density is zero everywhere else.

8. The Patterson method works by using a trial structure to calculate relative phase angles.

9. The Patterson function is also a Fourier series that takes the squared values of the structure factors.

10. Peaks in the Patterson map represent the vector between the heaviest atoms in the crystal structure.

Macromolecules – structure solution

11. The phase problem for macromolecules can be solved with direct methods, isomorphous replacement (IR) or multi-wavelength anomalous diffraction (MAD).

12. There are two direct method approaches for macromolecules:

    (a) Direct methods similar to those used for small molecules;

    (b) A shake-and-bake approach.

13. Isomorphous replacement involves the use of phases from known solved structures as a model or as phase estimates for a new protein crystal.

14. Crystals from several heavy metal derivatives of the same protein may be collected to enhance the use of IR.

15. Multi-wavelength anomalous diffraction involves the use of multiple X-ray wavelengths to collect data from one crystal.

16. Multi-wavelength anomalous diffraction exploits the anomalous scattering of a crystal lattice and uses X-ray wavelengths near the absorption edges of the heavy metal content in a crystal lattice to enhance this anomalous scattering.

17. Once a crystal structure is solved, the structure is then taken through to the process of refinement.

📖 FURTHER READING

Drenth, J. (1999). *Principles of Protein X-ray Crystallography,* 2nd edn. Springer Advanced Texts in Chemistry. Springer-Verlag, New York.

Giacovazzo, C., Monaco, H. L., Artioli, G., *et al.* (eds.) (2002). *Fundamentals of Crystallography,* 2nd edn. IUCr Texts on Crystallography. Oxford University Press, New York.

Glusker, J. P., Lewis, M., and Rossi, M. (1994). *Crystal Structure Analysis for Chemists and Biologists.* VCH Publishers, New York.

Rhodes, G. (2006) *Crystallography Made Crystal Clear: A Guide for Users of Macromolecular Models,* 3rd edn. Academic Press, London.

🔗 LINKS

Online learning tools

A Java programme to help understand Fourier Transforms: http://www.falstad.com/fourier/

Kevin Cowtan's Book of Fourier: **http://www.ysbl.york.ac.uk/~cowtan/fourier/fourier.html**

REFINING CRYSTAL DATA   7

By the end of this chapter you should be able to:

- Understand the least squares refinement process;
- Describe the parameters involved in the refinement process;
- Outline the challenges that may occur in the refinement of crystal structures;
- Explain crystalline state interactions such as hydrogen bonding and π-π interaction.

7.1 INTRODUCTION

Structure solution and resolving the phase problem produces an electron density map in which atoms are named and placed to form a recognizable model of the molecule in the crystal lattice.

Once a structural model is obtained from the structure solution process, it becomes necessary to compare the structure solution against the experimental diffraction data to see how well it 'fits'.

Within the structural model, however, the positions of the atoms and the associated displacement parameters (explained in Section 7.2.2) of the atoms are preliminary and the precision and accuracy of these parameters are then further enhanced through a cyclic process. This process of 'fitting' the model against the experimental diffraction data is known as *refinement*.

7.2 REFINEMENT OF CRYSTAL STRUCTURES

Section learning outcomes

To be able to:

- Understand and describe the refinement process;
- Identify the parameters involved in refinement;
- Explain how the parameters change over the course of the refinement;

- Understand and outline the use of constraints and restraints;
- Outline how hydrogen atoms are added to the final structural model;
- Compare experimental data collection at different temperatures.

Remember that the experimentally obtained intensity is a function of the observed structure factors
$$I_{hkl}^{corr} \propto (F_{hkl}^{obs})^2$$

Refinement refers to an iterative process, in which the molecular model is continually fitted and compared against the experimental dataset allowing very slight changes of movement (positional coordinates) and atomic displacement parameters (explained further in Section 7.2.2). The changes enable a better agreement between the calculated structure factors (structural model) and observed structure factors (experimental data).

The refinement of crystal structures is based on a statistical method, which can provide the crystallographer with a numerical measure of how well the model fits with the experimental data. This statistical measure is known as the least squares method.

7.2.1 The least squares method

The least squares method is often used as a statistical tool to compare with a specific value of certainty that a solution, or model, is agreeable with the experimentally obtained data. In short, the least squares method is a method of statistical analysis to provide numerical evidence as to how well the structural model (obtained from the structure solution process) fits the experimental diffraction data.

The least squares method is based on the assumption that there is usually a high ratio of experimental observations to defined parameters and that the experimental errors within the data follow the Gaussian or normal distribution.

A Gaussian or normal distribution is a bell-shaped curve, where the data distribution, x, is clustered around the central peak value of the mean, μ. The values on either side of the mean represent the value of error $(x - \mu)$. The curve shows that if the value of $(x - \mu)$ is small, it is more likely to be closer to the mean.

The curve can be mathematically represented by the equation:

$$N = \frac{1}{\sqrt{2\pi}} e^{-\frac{1}{2}(x-\mu)^2}.$$

Another method of representing the value of error is by the value of variance, σ^2, which also describes the distribution of the curve. The square root value of the variance, σ, is the *standard deviation* and is also known as the estimated standard deviation (e.s.d.). By using the values of the σ, the Gaussian distribution allows us to be able to quantify how different two values of measurements are.

Based on the curve below, we can say confidently that it is likely that there is a 99% probability that a measurement will be within three times the e.s.d. (3.0σ); alternatively we can say that if two measurements lie within 3σ of the mean there is a 99% probability that the two are the same, within the error limit.

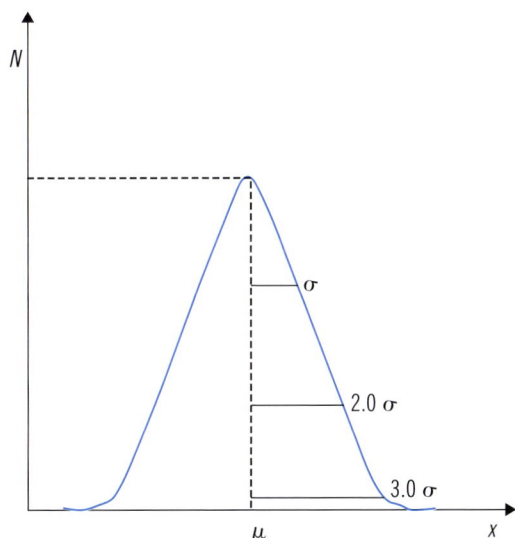

In crystallography, the least squares method of analysis provides the crystallographer with a numerical value to define the 'goodness', which is the accuracy of a model. The iterative refinement process that takes place continually tweaks the positions of the atoms and the related atomic displacement parameters (see Section 7.2.2) of the calculated model, to provide a better 'fit' with the experimental data. In essence, during the refinement process we are trying to minimize the difference between the observed structure factors and the calculated structure factors, $(|F_o| - |F_c|)$.

$$\Phi = \sum_{hkl} \omega_{hkl} \left(|F_o| - |F_c| \right)^2_{hkl}$$

Φ is used to represent the sum of the squares of the differences between the observed and the calculated structure factors. Each of these values is multiplied by a weighting factor, ω_{hkl}. It is the aim of the refinement process to minimize Φ.

The refinement process is monitored in the least squares method by a residual index, known as the R-value or R-factor. By mathematical definition, the R-value, R_1 is

$$R_1 = \frac{\sum \left| |F^{calc}| - |F^{obs}| \right|}{\sum |F^{obs}|}.$$

The R-value is the fraction determined by the summation of all the differences between the calculated and observed structure factors divided by the observed structure factors. This is often used as a 'yardstick' of a refinement process. Generally, the lower the R-value, the better the fit between the model and the experimental data.

Finding the minima: graph of R value vs changes in parameters

The graph in Fig. 7.1 provides a snapshot of the refinement process, in which the R-value (y-axis) changes during the course of a refinement, based on the changes in the parameters

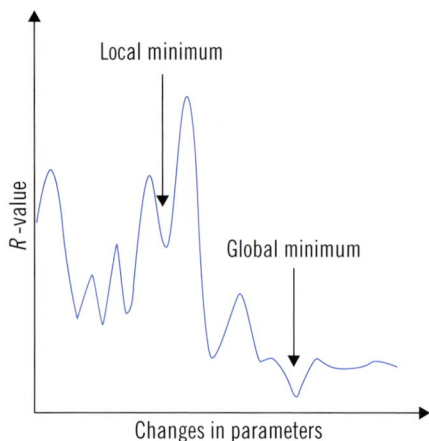

FIGURE 7.1 Minima in the least squares refinement

(*x*-axis). We find that there are several possibilities of the *R*-value falling into a local energy minimum. It is important that the model is as accurate and as close to the true structure as possible, to drive the refinement process into the global energy minimum. However, if the structure is not accurate, it is possible that the refinement process may trap the model in a local energy minimum, which is then not the true structure.

We can only hope to minimize such occurrences by ensuring that the model is as close to the true structure as possible and by monitoring the *R*-value and the various shifts in the atomic parameters (fractional coordinates) and displacement parameters of the structure.

7.2.2 Fractional coordinates and atomic displacement parameters

During the course of refinement, the atoms within the model are continuously moved very slightly, to find the overall energy minima. Each atom in the model can be defined by three positional parameters that are its coordinate-based location, as well as an overall displacement (thermal) parameter.

Through the structure solution process, the atoms are located within the asymmetric unit. The locations for each atom are described in terms of fractional coordinates, that is, the position of each atom is reported as a fraction of the length of the cell axes along which its position is described.

For example, Table 7.1 is an extract from a file (a *.res* file) undergoing a crystallographic refinement. It shows how the atoms are described in terms of fractional coordinates.

The column defining atom type lists the atoms by name on the left, and the numbers that assign the atom types on the right.

If we consider the iron atom Fe1, it is described by its position at *x* as 0.796501. This can be described in real terms as being 79.6501% of the length of the *a*-axis. It is similarly determined for positions along *y* and *z*.

The column defining occupancy also tells us whether or not the atom lies in a special position (refer to Section 4.4). When an atom has a full occupancy, that is, there is no higher symmetry influencing the atom, it is described (in the *.ins* and *.res* files) by 11.0000.

The extensions of computer file names used in this text, such as *.hkl*, *.res*, and *.ins*, refer to those used by the SHELX suite of programs.

TABLE 7.1 Atoms in a refinement process

Atom type		x	y	z	Occupancy	Temperature factor
Fe1	5	0.796501	1.000000	1.000000	10.50000	0.01831
O1	3	0.743194	0.965644	0.863820	11.00000	0.02610
O2	3	0.913087	0.928428	0.887423	11.00000	0.02436
O3	3	0.535391	0.819751	0.931587	11.00000	0.02015
O4	3	0.664254	0.852220	0.984754	11.00000	0.02325
C1	1	0.894974	0.922914	1.037234	11.00000	0.02274
C2	1	0.846979	0.874633	1.002080	11.00000	0.02756

However, if the atom sits in a special position related by symmetry, the occupancy is shown here. For example, the iron atom Fe1 in the position of (0.796501, 1.000000, 1.000000) has an occupancy of 10.50000; this means that within the asymmetric unit, the content of Fe1 is only half, and this will map onto itself in another asymmetric unit.

The final column in Table 7.1 lists the atomic displacement parameters. These are also known as thermal ellipsoids, thermal parameters, temperature factors, or vibration parameters.

All atoms vibrate and their atomic displacement parameters provide evidence of these vibrations. In the initial structure solution stage, these are not visible as the solution does not take full account of these vibrations; however, through the refinement process, these displacement parameters are factored into the structural model.

The initial structural model has atoms with isotropic displacement parameters, that is, with one value of the displacement parameter. The atoms are visually represented as round spheres. These provide an idea of the best fit of electron density to the atom type. For example, if the value of the displacement parameter is too large, the atom should be of a lower atomic number, while if a number is too low, an atom with greater electron density is required to fill it.

It is often difficult to pinpoint for certain how large or how small a value should or should not be, as consideration has to be given to the structure as a whole. However, as a general rule, it is often common to note that a displacement parameter is considered high when it approaches 0.1. When the experimental electron density is much higher than that of the assigned atom type, we find that the displacement parameter becomes very close to 0.00001. This is a clear pointer that an atom of a higher atomic number must be placed in that position.

In the course of refinement, all of the non-hydrogen atoms are then made anisotropic, which means that further refinement of the displacement parameters takes place. Instead of just one displacement parameter for each atom, once the atom is redescribed as anisotropic, each atom is then refined against six values, corresponding to the major axes of the ellipsoid and the orientational relationship to the reciprocal cell axes of the crystal. Anisotropic refinement allows for better modelling of the thermal vibrations within each atom and subsequently describes the ellipsoidal shape of the atoms.

```
Fe1  5 0.796501  11.000000  11.000000  10.50000  0.01831  0.03339 =
     0.03351   0.00507   0.00000   0.00000
O1   3  0.743194  0.965644  0.863820  11.00000  0.02610  0.04526 =
     0.03906   0.00199   0.00123   0.00844
O2   3  0.913087  0.928428  0.887423  11.00000  0.02436  0.04205 =
     0.04276  −0.00436  −0.00285   0.00240
O3   3  0.664254  0.819751  0.931587  11.00000  0.02015  0.04895 =
     0.03812   0.00022  −0.00363  −0.00405
O4   3  0.535391  0.852220  0.984754  11.00000  0.02325  0.04613 =
     0.03319  −0.00008  −0.00385  −0.00523
C1   1  0.894974  0.922914  1.037234  11.00000  0.02274  0.03465 =
     0.04272   0.00659  −0.00482   0.00085
C2   1  0.846979  0.874633  1.002080  11.00000  0.02756  0.02522 =
     0.04167   0.00085  −0.00360  −0.00123
```

FIGURE 7.2 Extract of part a .*res* file

In an extract of a .*res* file shown in Fig. 7.2, the six anisotropic temperature factors for the atom Fe1 are the final six numbers listed (highlighted in blue) relating to each atom.

7.2.3 Estimated standard deviations

Estimated standard deviations (e.s.d.) provide a measure of error for an experimentally obtained value. Following the structure solution and refinement process in a diffraction experiment, we are able to obtain a large number of numerical values including calculated and derived values, such as bond lengths and angles, and torsion angles.

It is for these values that the e.s.d. is able to provide an idea of how accurate these measures are. An e.s.d. is usually written to the value of one significant figure, and is written in brackets following a refined parameter. For example, a bond length of 1.5423 Å with a value of one (1) e.s.d. of ±0.0004 Å is written as 1.5423(4) Å.

From a practical and applied perspective, it is also possible to surmise that the e.s.d. defines the measure to be within the region of $(+x)$ or $(-x)$ *amount* from the stated measurement. As errors in crystallography are usually calculated to a value of three e.s.d.s (3σ), a bond length of 1.5423(4) Å also means that the bond length (within its errors) ranges from 1.5411 Å to 1.5435 Å.

7.2.4 Constraints and restraints

Constraints and restraints are methods with which certain parameters can be fixed to certain values, and these are particularly helpful for difficult refinements.

A constraint is an exact mathematical relationship, which when applied ensures that not all the parameters can be freely or independently refined. Constraints are absolutely rigid and therefore must be used with care to ensure that they are not inconsistent with the diffraction data. Unsuitable constraints will force other parameters to compensate and thereby introduce inaccuracies to the overall structure. Examples of constraints most commonly used are (a) atom site occupancy factors, which, in a disordered structure (see Section 7.3.1) sum to unity and (b) the refinement of hydrogen atoms (see Section 7.2.5), in which a 'riding model' is used to position the hydrogen atoms in an idealized geometry.

A restraint, however, is an approximate target value for a particular parameter, which can provide some extra information about a structure. It can be treated as an experimental observation and can be used with a weighting scheme applied to determine how closely the target value needs to be met. Restraints are most often applied to geometry and to atomic displacement parameters.

7.2.5 Hydrogen atoms

All initial molecular models derived from the structure solution process contain only non-hydrogen atoms. This is because the hydrogen atom is unable to scatter X-rays efficiently and cannot be located accurately within the electron density map.

Once the non-hydrogen atoms are accurately located and refined anisotropically, we are able to 'fix' the hydrogen atoms onto the relevant atoms, in what is usually known as a riding model. This means that the relevant hydrogen atoms are modelled onto the parent atom to which they are to be bound. These hydrogen atoms are subject to the same refinement process, in which their positions are allowed to move slightly within the confines of the defined modelling parameters. These hydrogen atoms are then provided with a displacement parameter, which is a percentage of the parent atom's displacement parameter.

With the development of more efficient diffractometers and better data processing software, it is now possible to locate hydrogen atoms (in some small organic molecules that diffract well) from the electron density difference map and subsequently refine them freely (with isotropic displacement parameters).

7.2.6 Collecting data at low temperatures

The temperature at which diffraction data from a crystal is collected can greatly influence the quality of the data and subsequently the precision and accuracy of the structural model solution. If data collection is carried out at room temperature, very often the displacement parameters of the atoms under study are larger because of the atomic energy vibrations. If we keep in mind that the X-ray beam also produces significant quantities of heat, the quality of the crystal (and the data) also diminishes more quickly during a routine data collection at room temperature.

Collecting crystal diffraction data at low temperatures helps to maintain crystallinity and reduce thermal vibrations in the atoms. This subsequently enables more accurate structure solutions and smaller anisotropic vibrations within atoms from better and more intense diffraction data.

The majority of low-temperature data collections are carried out with the use of a cryostat that passes a continuous stream of cold gas to shroud the experimental crystal. The temperatures can be reduced to approximately 100 K using liquid nitrogen, or to the order of 10 K with the use of helium.

While low temperatures help increase the overall quality of the diffraction data and subsequently enhance structure solutions, there remain some challenges, such as disorder and twinning, which occur occasionally and need to be overcome.

When these challenges have been overcome, how are we able to decide when a structural refinement is 'finished'? The overall end point of the refinement process is not as defined as a 'full stop' but tends to be a gradual process in which convergence is reached.

7.2.7 Refinement to convergence

Throughout the refinement process, minor movement and adjustments are continuously made to the structural model. The 'end point' to the refinement process is known as convergence. This is when the structural model is found to be in its best fit to the experimental data.

Convergence is said to occur when the overall energy minimum of the structure is found. This means that within the structural model the movements are very, very small, or almost non-existent. When the calculated and observed structure factors converge, theoretically there is almost no difference between the structural model and the actual molecule within the crystal lattice (obtained from the diffraction pattern).

At this point in the refinement process, there should usually be no significant (chemically sensible) residual electron density and atomic shifts between adjacent refinement cycles should be close to zero.

A good overall measure of this is the R-factor (see Section 9.2.1). Convergence will usually lead to the lowest stable R-values. Within the refinement process, there are also other indicators of convergence that are determined by the structural shifts, which are defined numerically.

SELF-TEST QUESTIONS

1. Refinement is a process that takes place after a structural model has been obtained.

   (a) What are the parameters that can be refined?

   (b) Describe the changes to these parameters that occur over the course of the refinement.

   (c) How do we know when the refinement process is complete?

2. Explain what is meant by the least squares method of refinement.

3. Briefly describe how errors are quantified.

4. What are the advantages of collecting data at low temperatures and how can this be carried out?

7.3 SOME CHALLENGES IN CRYSTALLOGRAPHIC REFINEMENTS

Section learning outcomes

To be able to:

- Describe the occurrences of disorder and twinning;
- Explain how static and dynamic disorder may be differentiated;
- Outline the differences between merohedral and non-merohedral twinning.

The majority of small-molecule crystal structures are straightforward in their solutions and refinement. However, for the majority of macromolecular crystals and a small percentage of small molecule crystals there exist challenges, such as disorder and twinning, that need to be overcome in the modelling process.

7.3.1 Disorder

Disorder occurs when some of the atoms in the structure adopt different orientations in different unit cells in the crystal; subsequently, the model in the solved structure represents two or more orientations of the atoms that are observed. The percentage of each orientation may vary with temperature but, assuming that the atoms are present in full occupancy, the various disordered components will sum to unity.

Disorder is commonly modelled as part of the refinement process; to locate and identify the residual electron density that may be found on the electron density map after most of the crystal structure has been assigned. The modelling of disorder, if correctly done, also contributes to the lowering of the R-factor.

Disorder within a crystal lattice often looks like the partial displacement of some atoms of a molecule within the crystal lattice and most commonly occurs in molecules that are able to take on differing orientations or different conformers.

Within a crystal lattice, it is most common to find disorder in long alkyl side chains or in solvents or counterions of crystallization.

In Fig. 7.3, we can compare two perchlorate atoms, (a) ordered (on the left) and (b) disordered (on the right). The disordered perchlorate atom takes on two almost equal conformations, which can be defined in the labelling scheme as Cl(2)-O(1A)-O(2A)-O(3A)-O(4) and Cl(2)-O(1B)-O(2B)-O(3B)-O(4). Simply put, disorder occurs when one molecule takes on two, sometimes overlapping, positions.

Disorder can be broadly divided into two categories: static, and vibrational or dynamic disorder.

Static disorder

Static disorder can occur when there are two conformers of the same molecule within the same unit cell or the conformers face in one direction in some unit cells and in another direction in other unit cells.

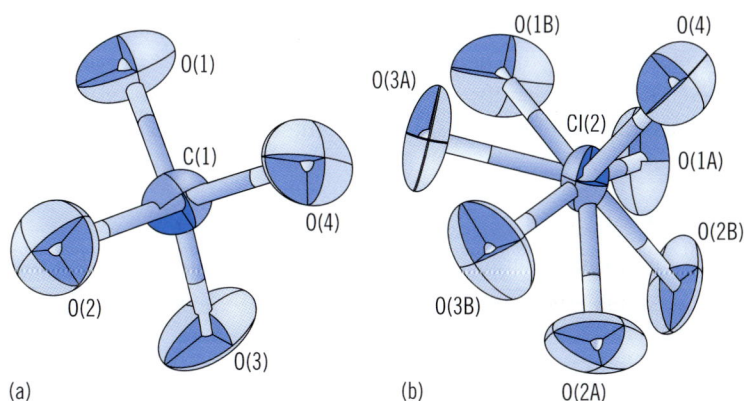

FIGURE 7.3 Two perchlorate molecules, (a) ordered (left) and (b) disordered (right)

The distribution of the conformers is not necessarily always an equal 50%:50% split. The occupancy of each disorder conformer can usually be determined, although not in a rigorous manner. The amount of occupancy for each conformer is usually determined based on how the electron density is distributed.

Vibrational or dynamic disorder

Vibrational or dynamic disorder occurs when the molecule is constantly undergoing flux. A typical example is that of ferrocene. This molecule undergoes a phase transition at 164 K, where close to this temperature the ring orientations of the ferrocene molecule are in constant motion. If the temperature is significantly above or below this temperature the ferrocene atoms are constant. Data from crystals containing ferrocene are usually collected either at 180 K or 120 K to avoid this temperature-related phase transition.

7.3.2 Twinning

Twinning is another challenging phenomenon that occurs in crystallography. Twinning often generally summarizes the situation in which one dataset contains diffraction from two or more crystals. Technically, however, a split crystal relates to two (or more) crystallites that are stuck together in different orientations, whereas a twinned crystal usually involves interpenetrating crystal lattices caused by faults in a single crystal. Both would result in 'twinned' diffraction data.

Finding a structure solution from a dataset of a twinned crystal is challenging and it is often easier to ensure that a crystal is single prior to collecting diffraction data.

As a preliminary means of identifying twinned crystals, it is routine simply to check the crystals visually under a polarizer on a microscope. Under a polarizer, single crystals will extinguish light, or become completely dark at certain positions when they are rotated, however if the crystal is twinned, only some parts of the crystal will extinguish light, while other parts remain bright.

Twinning can broadly be divided into two categories, merohedral and non-merohedral.

Merohedral twinning

Of the two types of twinning, merohedral twinning is the harder to identify. In essence, a merohedral twin occurs when two crystal lattices are connected in such a way that from the collected diffraction data, it appears to be single crystal. The diffraction data of the twinned crystal consequently belong to a higher symmetry than should be the case. This form of twinning is often harder to identify, as it is sometimes possible to solve and even refine the structure in this incorrect higher space group, although disorder is commonly found in such structures.

Non-merohedral twinning

Non-merohedral twins are usually easier to recognize, as the diffraction data are not easily resolved or there are clear signs of more than one reciprocal lattice, due to overlapping spots in the diffraction pattern.

In both cases, it is possible to resolve the twin once the relationship between the two individual crystals can be identified. This is usually a mathematical relationship, known as the twin law, and it defines how both lattices are related in three dimensions.

7.3.3 Polymorphism

A polymorph has been defined as, 'A solid crystalline phase of a given compound resulting from the possibility of at least two different arrangements of the molecules of that compound in the solid state.'[1]

Polymorphism is known simply as the ability of a molecule to exist in more than one crystalline form. It is particularly common in organic compounds, where different packing possibilities exist from the variations in intermolecular hydrogen bonding (see Section 7.4.1). Owing to the variation in the physical arrangement of the molecules, the related polymorphs have different physical properties, such as melting point, solubility, hardness, and density.[2] The most well known form of polymorphism is probably that of the element of carbon – resulting in graphite or diamond.

Polymorphism is believed to be influenced by the crystallization process. Although not much is known about the process of crystallization, two key stages have been identified: (a) the formation of a nucleus and (b) the subsequent growth of the crystal.[1] The first stage, which is the formation of the least stable product, is also associated with a free energy of activation;[2] this then finishes with the formation of the most stable product, the specific polymorph.[3]

While polymorphism does not usually cause problems in the refinement stages of crystallography, it is often useful to check whether a polymorph has previously been determined. This is often quickly and conveniently done by checking the parameters of the unit cell (in the initial stages of data collection) against a crystallographic database (outlined in Chapter 9).

An example of a polymorphic crystal is the molecule 2,2':6',2"-terpyridine (see Fig. 7.4).

Terpyridine crystallizes in two polymorphic forms. One form crystallizes in the *orthorhombic* space group $P\,2_1\,2_1\,2_1$ with one molecule in the asymmetric unit while the other crystallizes in the *monoclinic* space group $P\,2_1/c$, with two molecules in the asymmetric unit.

The orthorhombic form was first documented by Bessel *et al.*[4] in 1992, while the monoclinic form was discovered by Cole *et al.* in 2005.[5] As one might expect, both polymorphs have very different packing arrangements.

The *monoclinic* form of terpyridine, as shown in Fig. 7.5, has a sandwich-herringbone motif, which consists of pairs of molecules parallel to each other and perpendicular to

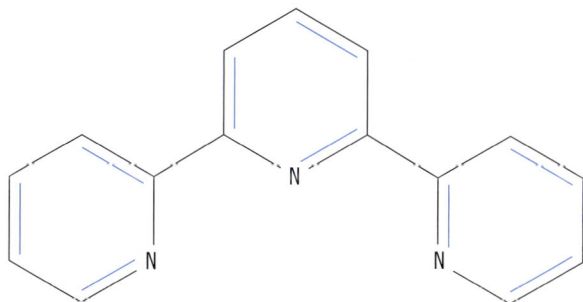

FIGURE 7.4 The structure of 2,2':6',2"-terpyridine

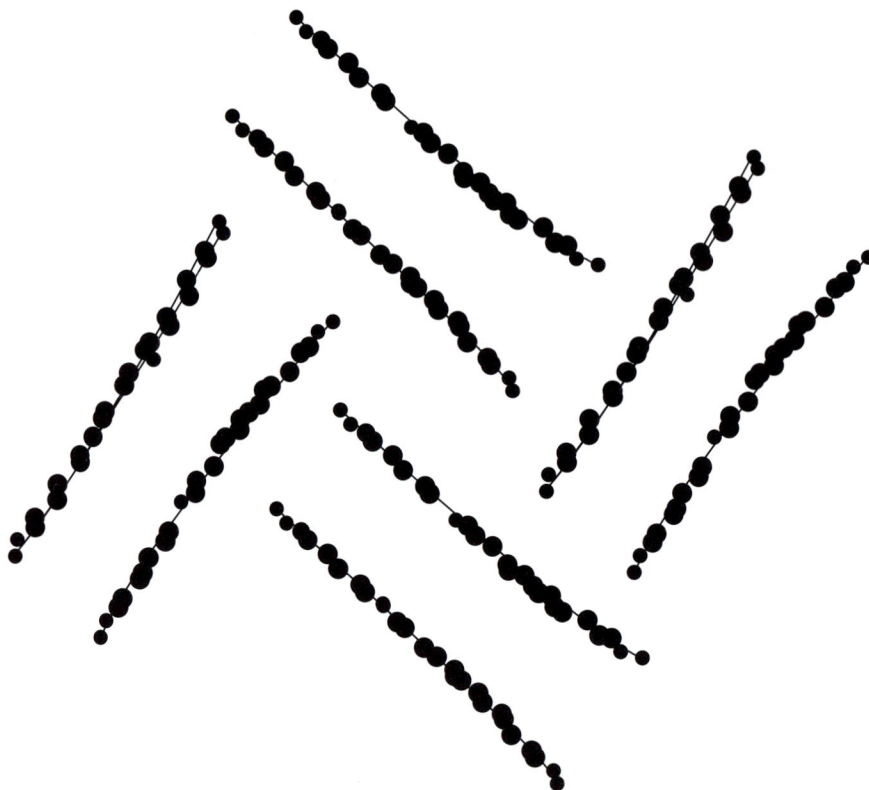

FIGURE 7.5 Packing of *monoclinic* terpyridine

another pair of molecules; i.e., the original herringbone motif is made up of sandwich-type diads (see Section 7.4.2 on π-stacking).

In the *orthorhombic* form, however, the asymmetric unit contains a very short axis of $a = 3.9470(10)$ Å. In Fig. 7.6, the crystal packing is viewed looking down the a-axis. The packing type is typical of a β-stacking motif, where the molecules are stacked at a distance of 3.947 Å down the a-axis (refer also to Section 7.4.2 on π-stacking).

SELF-TEST QUESTIONS

1. Describe what is meant by the term *disorder in* crystallography.

2. With the use of examples, compare the types of disorder that can occur.

3. Outline the differences between merohedral and non-merohedral twinning.

4. What is meant by the term *polymorphism?* Citing an example, explain how the characteristics of polymorphs may differ.

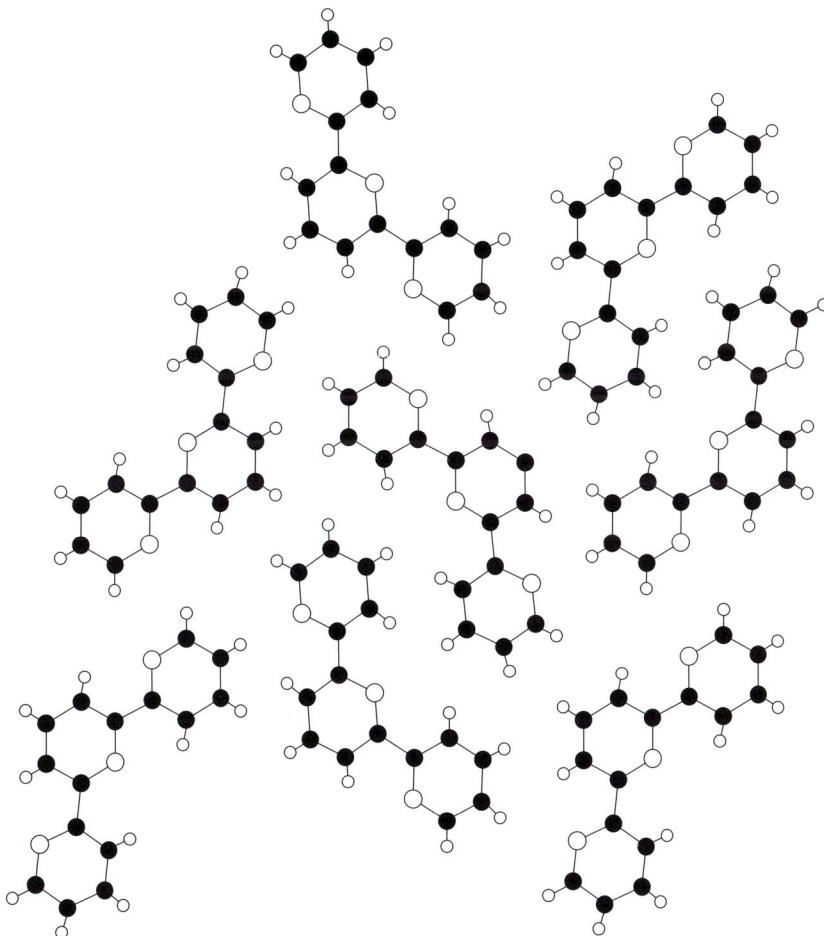

FIGURE 7.6 Packing of *orthorhombic* terpyridine

Reproduced by permission of the Royal Society of Chemistry from: Bessel, C. A., See, R. F., Jameson, D. L., Churchill, M. R., and Takeuchi, K. J. (1992) Structural considerations of terdentate ligands: crystal structures of 2,2:6,2-terpyridine and 2,6-bis(pyrazol-1-yl)pyridine. *Journal of the Chemical Society: Dalton Transactions*, (22):3223.

7.4 INTERACTIONS IN THE CRYSTALLINE STATE

Section learning outcomes

To be able to:

- Outline and describe the different solid state interactions;
- Recognize and identify types of hydrogen bonds and π-π interactions;
- Compare the types of hydrogen bond in terms of lengths and angles;
- Describe types of π-stacking that can occur.

We have been mainly concerned so far with the identification of molecular content within a crystal lattice. However, on completion of a structure solution, it is possible to garner a whole host of information relating to the bonds and bond angles within the molecule (intramolecular bonding), intermolecular interactions, and other interactions, such as hydrogen bonding and π (pi)-stacking, as detailed next.

7.4.1 Hydrogen bonding

Hydrogen bonds[6,7] are non-covalent interactions that specifically involve donor–acceptor interactions, where the donor is a coordinated hydrogen atom. In biological molecules, hydrogen bonding plays a significant role. Some of these roles include holding two strands of DNA together, helping enzymes bind to substrates, and working to bind antibodies to antigens.

Hydrogen-bond energies range from 15–40 kcal/mol^{-1} for strong bonds, 4–15 kcal/mol^{-1} for medium bonds, and 1–4 kcal/mol^{-1} for weak bonds. Strong hydrogen bonds, also known as 'ionic hydrogen bonds' are often formed in groups where there is a deficiency of electron density in the donor group (A–H), O$^+$–H, N$^+$–H, or an excess of electron density in the acceptor group (B), F$^-$, O$^-$–H, O–C, O–P or N$^-$.

Meanwhile, moderate hydrogen bonds are mostly formed by neutral donor and acceptor groups, such as O–H, N–H, and O=C, where donor atoms (A–H) are electronegative relative to the hydrogen and the acceptor, B, has lone-pair unshared electrons. These 'normal' hydrogen bonds form important functions and are essential components of biological systems.

Weak hydrogen bonds are formed when the hydrogen atom is covalently bonded to an atom that is slightly more electroneutral relative to hydrogen, for example, C–H, Si–H. Weak hydrogen bonding also occurs when the acceptor group has no lone pairs but has π electrons, such as C≡C or an aromatic system. These interactions are similar in energy and geometry to those of van der Waals complexes but can be differentiated from them by the direction of the A–H bond.

Quantities used to measure the geometry of the hydrogen bond are: (a) the A–H covalent bond length, (b) the H---B hydrogen bond length, and (c) the A---B hydrogen bond distance. A strong hydrogen bond would have a shorter A---B distance, while weaker hydrogen bonds would have longer A---B bond distances. The bond angle is also defined by these parameters that is, the A–H---B bond angles. It is only in strong hydrogen bonds that the angle is close to 180°; in moderate and weak hydrogen bonds, this angle deviates from linearity. Moderate and weak bonds have a wide spread of hydrogen bond lengths and bond angles in the crystalline state, and also show a compromise with other packing forces.

Hydrogen bonds can be calculated as part of the computer-based refinement process. This is then usually confirmed with a visual examination of the listed hydrogen bonds. Fig. 7.7 shows some examples of hydrogen bonding.

The hydrogen bond shown in Fig. 7.7(a) can be described by the following parameters: the hydrogen bond has a distance of O(4)–H(4)---F(15) = 2.522(4) Å and an angle of 156.7(1)°. Similarly the hydrogen bond in Fig. 7.7(b) can be said to have a distance of N(1)–H(1)---O(1) = 2.560(3) Å and an angle of 158.7(3)°.

7.4.2 π–π interactions

π–π interactions are also known as π-stacking. These are interactions in the crystalline state that occur between two (or more) adjacent aromatic phenyl rings. These π–π interactions

(a)

(b)

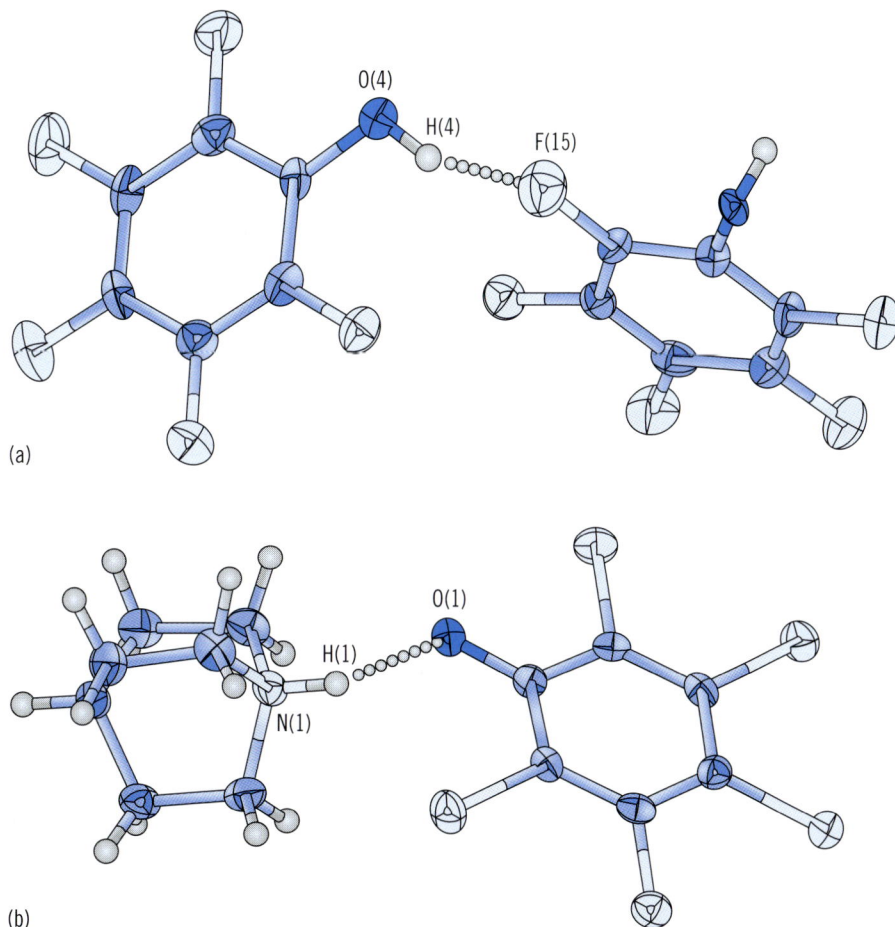

FIGURE 7.7 Some examples of hydrogen bonding

occur when attractive interactions between π-electrons and the σ-framework outweigh unfavourable contributions, such as π-electron repulsions. It has been shown that although π–π interactions are controlled by electrostatic interactions, the major energetic contribution comes from other factors.[8]

Figure 7.8 schematically represents a carbon atom in a π-system. Each carbon atom in the π-system is considered to have a charge of +1 at the nucleus of the atom and two $-\frac{1}{2}$ charges at a distance, δ, above and below the plane of the π-system. Generally, π–π geometries are either repulsive or attractive.

Figure 7.9 represents the types of geometry that are most commonly found for π–π interactions. The diagram in Fig. 7.9(b) represents a face-to-face geometry in which the repulsive forces are evident; however, if one π-atom is rotated through 90° relative to the other and one π-atom is offset laterally relative to the other, there is an attraction band.

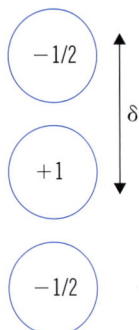

FIGURE 7.8 Model for an atom contributing one electron to the molecular π-system

Adapted with permission from: Hunter, C. A. and Sanders, J. K. M. (1990). The nature of π–π interactions. *Journal of the American Chemical Society* **112**:5525–34. ©1990, the American Chemical Society.

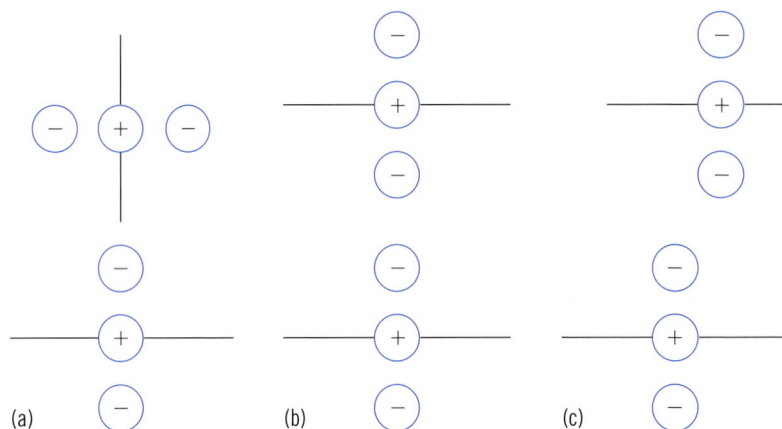

(a)   (b)   (c)

FIGURE 7.9 (a) The edge-on or T-shaped geometry; (b) the face-to-face geometry; (c) the offset π-stacked geometry.

Adapted with permission from: Hunter, C. A. and Sanders, J. K. M. (1990). The nature of π–π interactions. *Journal of the American Chemical Society* **112**:5525–34. ©1990, the American Chemical Society.

In Fig. 7.9(a), rotations between 0° and 90° can lead to attraction at small offsets in an edge-on arrangement. However, for rotations between 90° and 180°, leading to repulsions at small offsets, this is again a face-to-face arrangement, as shown in (c).

In essence, π–π repulsion dominates face-to-face stacked geometry while π–σ attraction dominates in edge-on or T-shaped geometries, while the π–σ attraction also dominates in an offset π-stacked geometry.[8]

Apart from individual interactive motifs, aromatic compounds also form four basic types of stacking pattern. These are (a) the herringbone motif, where carbon–carbon (C⋯C) non-bonded interactions are between non-parallel neighbouring molecules; (b) the sandwich-herringbone, which is made up of sandwich-type diads; (c) the γ-motif, in which the main C⋯C interactions are between parallel translated molecules; and (d) the β-motif, which is made up of graphitic planes characterized by strong C⋯C

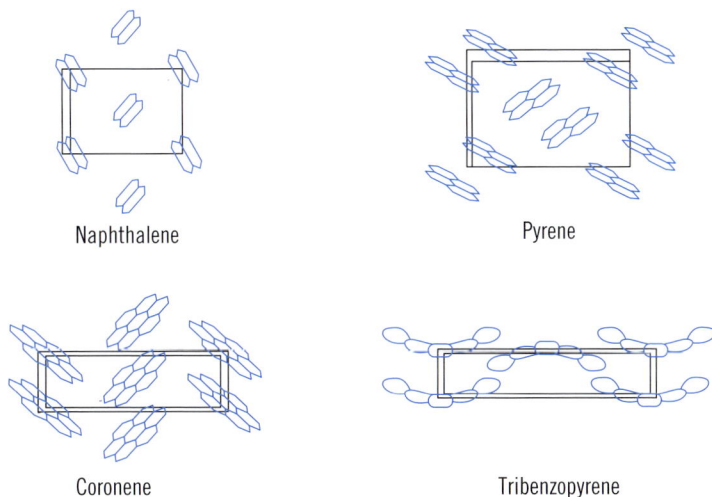

FIGURE 7.10 The four basic aromatic crystal packing, as exemplified by naphthalene (herringbone), coronene (γ), pyrene (sandwich-herringbone), and tribenzopyrene (β).

Reproduced with permission of the International Union of Crystallography from: Desiraju, G. R. and Gavezzotti, A. (1989). Crystal structures of polynuclear aromatic hydrocarbons. Classification, rationalization and prediction from molecular structure. *Acta Crystallographica Section B: Structural Science* **B45**:473–82.

interactions.[9] Figure 7.10 shows graphical representations of the four types of stacking motif.

The description of π-stacking is usually in terms of centroid–centroid distance; this is where the distance is measured from the centres of the two interacting phenyl rings and the offset angle is based on the angle at which both the phenyl rings are offset.

Unlike hydrogen bonds, which can be calculated, π–π interactions are usually identified by examining the overall packing patterns within the structural model that contains phenyl rings. If π–π interactions are located, centroid–centroid distances and offset angles can be calculated.

SELF-TEST QUESTIONS

1. Hydrogen bonding and π–π interactions are the most common type of molecular interactions in crystal structures. Write a brief description of each.

2. Outline the types of hydrogen bonding that can occur and the elements most likely to be involved in each.

3 What types of molecule are involved in π–π interactions?

4. Briefly describe the types of geometry that can be found for π–π interactions.

5. List the four types of stacking pattern that can be found for π–π interactions.

▣ CHAPTER SUMMARY

1. Refinement is an iterative process based on a statistical measure (usually the least squares method), which is carried out after structure solution.

2. The purpose of refinement is to minimize the differences between the calculated structure factors obtained from the structure solution and the observed structure factors from experimental diffraction data.

3. During the course of refinement, the atomic coordinates and atomic displacement parameters are continuously moved, to locate an overall energy minimum for the crystal structure.

4. Estimated standard deviations (e.s.d.) are used as a measure of the error for parameters within a crystal structure.

5. Hydrogen atoms are usually added in a 'riding model' towards the end of the refinement process.

6. Collecting data at low temperatures improves the overall quality of the diffraction data.

7. Disorder (static and dynamic) and twinning (merohedral and non-merohedral) are two types of challenge that can occur in determining crystal structures.

8. Intramolecular and intermolecular interactions can occur in a crystal lattice.

9. The most common interactions are hydrogen bonding and $\pi-\pi$ interactions.

☰ REFERENCES

1. Dunitz, J. D. and Bernstein, J. (1995). Disappearing polymorphs. *Accounts of Chemical Research* **28**:193–200 (and references therein).

2. Cross, W. I., Blagden, N., Davey, R. J., *et al.* (2003). A whole output strategy for polymorph screening: combining crystal structure prediction, graph set analysis, and targeted crystallization experiments in the case of diflunisal. *Crystal Growth and Design* **3**:151–8.

3. Blagden, N. and Davey, R. J. (2003). Polymorph selection: challenges for the future? *Crystal Growth and Design* **3**:873–85.

4. Bessel, C. A., See, R. F., Jameson, D. L., Churchill, M. R., and Takeuchi, K. J. (1992). Structural considerations of terdentate ligands: crystal structures of 2,2:6,2-terpyridine and 2,6-bis(pyrazol-1-yl)pyridine. *Journal of the Chemical Society: Dalton Transactions* (22):3223.

5. Bowes, K. F., Clark, I. P., Cole, J. M., *et al.* (2005). A new polymorph of terpyridine: variable temperature X-ray diffraction studies and solid state photophysical properties. *CrystEngComm* **7**:269–75.

6. Jeffrey, G. A. (1997). *An Introduction to Hydrogen Bonding*. Topics in Physical Chemistry. Oxford University Press, New York.

7. Steed, J. W. and Atwood, J. L. (2000). *Supramolecular Chemistry*. John Wiley & Sons Ltd, Chichester.

8. Hunter, C. A. and Sanders, J. K. M. (1990). The nature of $\pi-\pi$ interactions. *Journal of the American Chemical Society* **112**:5525–34.

9. Desiraju, G. R. and Gavezzotti, A. (1989). Crystal structures of polynuclear aromatic hydrocarbons: classification, rationalization and prediction from molecular structure. *Acta Crystallographica Section B*: Structural Science **B45**:473–82.

📖 FURTHER READING

Giacovazzo, C., Monaco, H. L., Artioli, G., *et al.* (eds.) (2002). *Fundamentals of Crystallography*, 2nd edn. IUCr Texts on Crystallography. Oxford University Press, New York.

Ladd, M. F. C. and Palmer, R. A. (2003). *Structure Determination by X-ray crystallography*, 4th edn. Kluwer Academic, New York.

Müller, P., Herbst-Irmer, R., Spek, A. L., and Schneider, T. R. (2006). *Crystal Structure Refinement*: *A Crystallographer's Guide to SHELXL*. IUCr Texts on Crystallography. Oxford University Press, New York.

THE CRYSTALLOGRAPHIC EXPERIMENT

8

By the end of this chapter you should be able to:

- Identify and describe the main techniques used to grow good quality crystals;
- Outline the main methods of mounting a crystal on a diffractometer;
- Identify, describe, and discuss the main components of an X-ray diffractometer;
- Outline and explain the crystallographic procedure.

8.1 INTRODUCTION

Although the earlier chapters in this book have been dedicated to the fundamental principles of X-ray crystallography, the use of modern diffractometers now allows most routine crystal structures to be collected and solved with very little crystallographic experience or training.

While the advancement and development of diffractometers and related computer hardware and software have significantly eased the process of collecting and solving crystal structures, the challenge of obtaining good quality crystals suitable for an X-ray diffraction experiment remains.

In this chapter, we look briefly at the various techniques that may be employed to obtain good crystals and how these crystals may be mounted on an X-ray diffractometer. We will also discuss the main components that constitute an X-ray diffractometer. Finally, we will examine the crystallographic process, from mounting a suitable crystal to the results of an X-ray diffraction experiment.

8.2 GROWING CRYSTALS

Section learning outcomes

To be able to:

- Identify the most common techniques used in the crystallization process;
- Discuss and analyse which crystallization technique may be most suitable.

It has sometimes been said that crystal growing is a 'black art'. Growing crystals of good quality often requires patience and a degree of trial and error, in experimenting with solvents and a variety of environmental factors (changes in temperature or pressure). A number of factors (among them temperature, pressure, choice of solvent, level of saturation, and nucleation) contribute to crystal growth and the acquisition of good crystals often involves balancing these factors.

The main aim of a crystallization experiment is to obtain crystals that are clear, well defined and of a suitable size, ideally about 0.1–0.3 mm in each dimension. Crystals often grow best when a saturated solution is left undisturbed; it is not surprising then that the best crystals are often obtained when a crystallization experiment has been forgotten.

The techniques for growing small molecule crystals are often more straightforward than those for growing macromolecular crystals. Macromolecular crystals are most commonly grown by the hanging-drop technique, although it is now common to have high-throughput robotic systems that automate the crystallization of proteins.

8.2.1 Slow evaporation

Slow evaporation is the quickest and easiest crystallization process to set up: a saturated solution of the compound is left to evaporate slowly. A suitable solvent for the compound is chosen. While slow evaporation is commonly carried out on laboratory benchtops, it is also possible to use this technique combined with temperature changes, either cooling or warming of the solution. The evaporation of the solvent causes supersaturation to occur and subsequently crystals may form.

8.2.2 Layering of miscible solvents

The layering of miscible solvents is another technique that is commonly used in growing crystals. The compound is dissolved in one solvent and then a non-solvent, which is miscible with the first solvent, but in which the compound does not dissolve, is gently placed above (or below) the solute. Crystals grow as the solvents diffuse into one another.

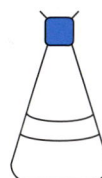

8.2.3 Vapour diffusion

Vapour diffusion has often been successful when other crystallization techniques have proved futile. Vapour diffusion occurs when the compound is dissolved in a small, open container and the small container is placed in a larger container with a volatile but miscible non-solvent. As the non-solvent diffuses into the solution, supersaturation occurs and crystals may form.

8.2.4 The hanging-drop technique

The hanging-drop technique is a variation of vapour diffusion and is most often used to grow protein crystals. In the hanging-drop method, a concentrated drop of the protein solution is suspended on a microscopic cover (usually silicone-treated) over a concentrated protein solution. As the solvent evaporates from the hanging drop, supersaturation can cause crystals to grow.

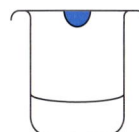

While there are many excellent texts outlining and discussing the methods most suitable for growing good quality crystals, there is no one definite method that can be prescribed. In cases when it may be difficult to grow good quality crystals, some trial and error may need to be used and sometimes unusual methods may yield the elusive crystals.

8.3 MOUNTING CRYSTALS

Section learning outcomes

To be able to:

- Identify the main techniques used to mount a crystal on the diffractometer;
- Discuss which crystal mounting technique may be suitable for specific types of compounds.

Another important aspect of collecting crystal data from a diffractometer is mounting the crystal on the diffractometer to ensure that it remains in place throughout the data collection process, and also to ensure that as much as possible is done to enhance the life of the crystal and protect the crystal from the heat generated by the X-rays.

8.3.1 Mounting on a glass fibre

A simple set-up for crystal mounting consists of a short length of glass fibre, usually smaller than the dimensions of the crystal sample, in which the crystal can be glued to the tip of the fibre. This works well for data collections at room temperatures; however, if the data collections are carried out at low temperatures, freezing of the glue under liquid nitrogen temperatures causes the crystals to drop off the end of the fibre.

An alternative to the use of glue is the use of an inert perfluoroether oil. The high viscosity of the oil enables the crystal to sit at the end of the glass fibre. This is then almost instantly frozen in place under the low temperatures of a cryostat. The layer of oil also serves to protect the crystal sample. This is a good method for protecting air-sensitive crystals.

8.3.2 Mounting in a glass capillary

Another technique that can be used to mount and collect data from air-sensitive crystals is to seal the crystal inside a glass capillary. This is sometimes necessary for extremely air-sensitive crystals.

8.3.3 Mounting in loops

The use of loops to mount crystals is common practice in protein crystallography. The protein crystal is usually mounted from a solution and surface tension maintains the position of the crystal within the loop.

The now common use of low temperatures during data collection means that the crystal and its surrounding solution is typically frozen in place once mounted on the diffractometer and during the data collection process.

SELF-TEST QUESTIONS

1. Describe two techniques commonly used in growing crystals

2. Discuss why it is important to the crystal structure determination process to obtain a good crystal.

3. Compare the crystal-mounting techniques that may be used to mount a crystal that is air-sensitive.

8.4 X-RAY DIFFRACTOMETERS

Section learning outcomes

To be able to:

- Outline and identify the main components of an X-ray diffractometer;
- Discuss and compare the different sources of X-rays;
- Describe the different types of detectors most commonly available.

The experimental instruments used to collect X-ray diffraction data are known as X-ray diffractometers. Several companies (among them Bruker AXS, Rigaku, Oxford Diffraction (Varian), and MAR Research) produce X-ray diffractometers for both commercial and academic use. While each diffractometer is specialized and may be slightly different in detail, the overall set-up of a diffractometer is standard.

Figure 8.1 shows a schematic representation of the general outline of a diffractometer. The X-ray diffractometer set-up typically consists of a detector, an X-ray source, a collimator, and a goniometer to hold the sample crystal. The variations that can be found in diffractometers are generally the types of X-ray source, types of detectors and slight differences in the overall arrangement. These differences are outlined below.

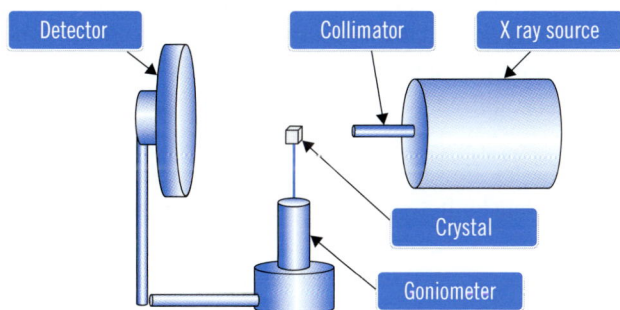

Detector   Collimator   X ray source

Crystal

Goniometer

FIGURE 8.1 Typical X-ray diffractometer set-up

8.4.1 X-ray sources

Typical sources of X-rays for diffractometers are based on X-ray generators. These are: (a) sealed X-ray tubes, (b) rotating anodes, (c) synchrotrons. Synchrotron radiation offers very-high-energy X-rays, which are often suitable for crystals whose structure cannot be determined on a laboratory diffractometer, although the diffractometer set-up is similar.

Sealed X-ray tube

In a sealed X-ray tube (vacuum tube), an electron beam is generated on a hot filament (cathode) using an electric current. The electron beam is then accelerated using a high voltage between the cathode and the anode. The anode is a metal target, typically Cu, Mo, Ag, Fe, Cr, or W (depending on wavelength requirements). The X-ray beam is produced as a by-product of the electrons colliding with the anode, upon which the electrons slow down.

As the continuous collision of electrons on to the anode causes it to be come very hot, it is usually cooled using a water source.

Rotating anode

Rotating anodes work on a similar premise to the sealed X-ray tube, with the main difference being that in order to reduce the heat the anode is made to rotate. This allows different parts of the anode to be used and also enables better distribution of the heating and cooling process. This also allows the voltage generated to be increased and consequently, the intensity of the X-rays is increased. X-rays from rotating anodes are between five and ten times brighter than the conventional sealed tube.

In both cases, the X-rays are then channelled out through beryllium (Be) windows towards a monochromator. The monochromator is usually a strong crystal (silicon or highly oriented graphite crystal) and is used to produce one wavelength of X-rays, based on Bragg's law.

For X-ray diffractometers with sealed tubes, a collimator is used to eliminate stray radiation. This pares down the X-ray beam using a tube with a small hole (usually between 0.35 mm and 0.5 mm) near to the X-ray source. On rotating anodes, a similar process is achieved with mirrors.

The resulting X-ray beam is aimed at a crystal that is centred so that it remains in the X-ray beam during the entire data collection process. The X-rays diffracted from the crystal are then collected on a detector.

Synchrotron radiation

Synchrotron radiation was first observed as a by-product of particle accelerations. When particles (electrons and positrons) are accelerated in magnetic fields, electromagnetic radiation, occurring over a wide range of the whole spectrum, is emitted. It was recognized that this electromagnetic radiation could be utilized for many spectroscopic techniques.

Initially, synchrotron radiation was exploited 'parasitically' from the particle accelerators designed for high energy physics, however, later, storage rings dedicated to the production of synchrotron radiation were built. Generally, storage rings make use of relativity to convert useless radio frequency energy into useful electromagnetic waves and X-rays through a Doppler shift.[1]

If we refer to Fig. 8.2, we see that a storage ring is essentially a circular arrangement of bending magnets (1.2 T or more), in which an electron (or positron) beam is made to move in a polygonal path. The electron beam consists of discrete bunches of electrons (not a continuous stream) injected into the storage ring from a linear accelerator. This electron beam then moves in a high vacuum pipe. However, as the pipe is an imperfect vacuum, with residual gas molecules, this causes a steady decline in synchrotron radiation intensity, leading the beam to need to be 'refilled' with electrons.

As the electrons move through the storage ring, the bending magnets have magnetic fields perpendicular to the plane of the ring. This causes electrons to change direction (subjected to an inward radial acceleration) as they pass through the bending magnets. Electromagnetic radiation (ranging from infrared to ultraviolet to the soft X-ray region) is thus produced at a tangent to the orbit of the electrons.

While the electromagnetic radiation emitted can be used for many spectroscopic techniques, crystallography requires X-rays of a higher energy than those produced by the storage ring. A wiggler (wavelength shifter) is used for the purposes of extending the energy range and increasing the intensity of the electromagnetic radiation emitted. A set of three magnets (of high fields, approximately 5–6 Tesla [T]) deflects the electron beam, sending it around a curve (a 'wiggle') and then returning it to its original path. This causes extra synchrotron radiation emission from the electron path to 'wiggle', thereby shifting it to a higher photon energy (shorter wavelength).

The use of X-ray sources from synchrotron radiation helps to overcome some of the challenges in crystallography. These challenges come mainly from the difficulty in obtaining a sample that diffracts well and often these limitations are caused by small poor-quality crystals. This problem is particularly acute in some areas of biological and materials science, where samples often include macromolecules, microporous solids, and supramolecular assemblies, where it is often extremely difficult to produce large enough crystals for a laboratory source of X-rays.

At the synchrotron, the higher flux (number of photons per second in a beam of radiation) and greater brightness and brilliance of the concentrated X-ray beam makes it possible to determine the structures of materials that give micron-sized crystals, which may not have produced an observable diffraction pattern with a laboratory source.

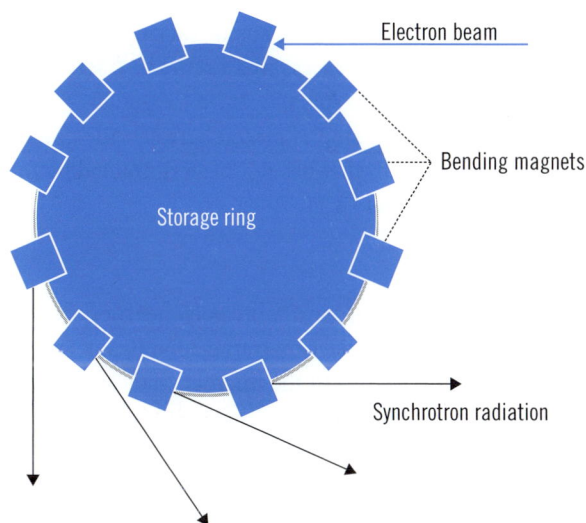

FIGURE 8.2 Schematic representation of a synchrotron storage ring

Other advantages of the synchrotron include the availability of a wide range of wavelengths that can be tuned using various monochromators. Silicon and germanium are often used as the standard monochromator, at which the wavelength is close to that of molybdenum radiation. However, the use of tuneable wavelengths is particularly advantageous for data collection using the MAD technique (Section 6.4.3).

8.4.2 Detectors

Historically, photographic film was commonly used to collect diffraction data. This required detailed identification of the crystal faces and alignment of the crystal in the incident X-ray beam. Much of the structure was determined by physically measuring and calculating the differences in the diffraction spots produced on the film.

Both the development of computers and modern-day imaging electronics mean that a standard modern diffractometer is now commonly equipped with either a charged-coupled device (CCD) or an image plate (IP).

Charged coupled device (CCD)

The technology for the CCD is most commonly known and applied in digital cameras. This is also the technology that has vastly improved the speed and accuracy at which the diffraction data from a crystal can now be collected and solved.

On an X-ray diffractometer, the CCD is located after the crystal, where it is able to 'collect information' from the diffracted X-rays. The CCD in essence consists of an array of pixels (these can be thought of as electron buckets that can be filled and emptied). The CCD is made of a layer of a type of semiconductor material (usually silicon), which is then covered with a transparent metal oxide film (of for example, indium tin oxide).

When the diffracted X-rays reach the CCD, the energy from the diffracted X-rays can liberate electrons from the surface of the CCD, which then creates positive holes. These holes are repelled by the already applied positive potential and move to the base, while the electrons are collected in a potential well. The stronger the source of X-rays, the more electrons are collected in the well.

This charge is then moved along channels (or rows), by means of varying potentials, towards the data-processing device. This is a very fast process that occurs in fractions of seconds, resulting in a readout of a 'frame' of diffraction spots.

When a CCD is used on a diffractometer, a beam stop is placed between the crystal sample and the CCD to stop the incident X-ray beam that has not been diffracted by the crystal, as the intensity of this beam could damage the CCD.

Image plate (IP)

An image plate is in essence a very flexible image sensor that consists of a polyester support film, coated with a layer of microcrystals of luminescent material (for example $BaFBr:Eu^{2+}$).[1]

When the diffracted X-rays reach the IP, the electrons in the microcrystals are excited and then become trapped in the lattice of the empty anion (F^-). This causes the colour of the luminescent material to change.

---

[1] Information from Fujifilm Imaging Plates: http://www.fujifilm.com/products/life_science/si_imagplate/whatis03.html

To collect this information for the luminescent material, a laser is used to scan the IP and the information channelled and transferred to a photomultiplier tube, where the information is converted into electric signals, which are processed and digitized to produce a 'frame' of diffraction spots.

Regardless of the type of detector used to collect diffraction data, it is always important to ensure that the best-quality crystal from the sample is selected and as much diffraction data required are collected. This refers to collecting what is known as 'sufficient redundancy', that is, while to determine the structure of an orthorhombic crystal one might only be required to collect data from one-eighth of the reciprocal lattice space, it is usually best practice to collect 'more' data so that there is a greater redundancy of data; perhaps a quarter or even a half of the reciprocal lattice space. (This is discussed further in Section 9.2.)

SELF-TEST QUESTIONS

1. Briefly outline the main components of an X-ray diffractometer.

2. Discuss and compare the different types of X-ray sources that may be used for X-ray data collection.

3. What are the types of detector that are most commonly used now? Briefly describe them.

8.5 THE CRYSTALLOGRAPHIC PROCEDURE

Section learning outcomes

To be able to:

- Outline and describe the crystallographic procedure;
- Relate the procedure to the background covered in the earlier chapters of this book.

So far, we have examined the individual techniques, processes, and other details that make up the entire crystallographic process, which, in the modern day, is almost entirely computer-driven.

The flowchart in Fig. 8.3 outlines the X-ray crystallographic procedure that takes place, from the identification of a suitable crystal on a microscope to the final submission of the crystal structure to one of the relevant databases.

At the start of the crystallographic process, prior to collecting data from a single crystal on the diffractometer, a sample of the crystals is examined under a light microscope and a polarizer.

8.5.1 Selecting a crystal (Chapter 8)

In examining the crystals under a microscope, we are looking for a clear single crystal that has clean faces and sharp edges. Typically, organic crystals would be about 0.3 mm^3, while

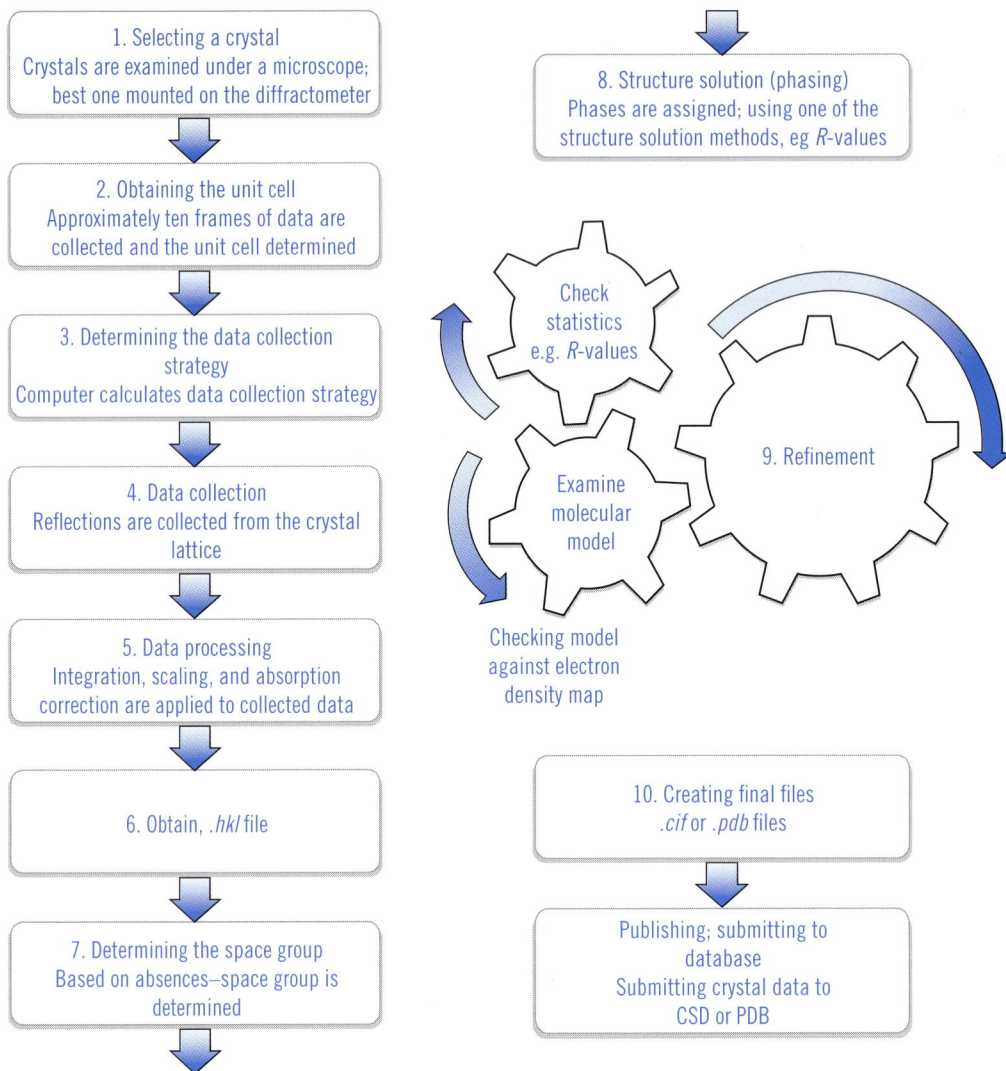

FIGURE 8.3 Flowchart detailing the crystallographic process

organometallic crystals can be smaller, as the heavy element within these crystals contributes to better scattering of the X-rays and consequently produces stronger reflections or diffraction spots. The crystals can be examined on a polarizer, to ensure that they are single. Once a crystal is selected, it is then mounted on the diffractometer. (Refer to Section 8.3.)

8.5.2 Obtaining the unit cell (Chapter 1)

When the crystal is mounted on the diffractometer, typically a quick scan, of approximately ten frames of data, is collected. This enables the computer to calculate the unit-cell dimensions of the crystal lattice and some statistical measures on how 'good' the crystal is.

8.5.3 Determining the data collection strategy

If the crystal is suitable for further data collection, the computer calculates the data collection strategy. This means that the computer would decide how much of the reciprocal lattice space needs to be collected to obtain sufficient data to solve and refine the structure.

For example, the high symmetry within the orthorhombic crystal lattice would require that only one-eighth of the reciprocal lattice space be collected, while a triclinic crystal may require half or even the whole reciprocal lattice space to be collected.

8.5.4 Data collection

Data collection then takes place. Depending on a number of factors, including the symmetry of the crystal lattice, how well the crystal sample diffracts, and the type of diffractometer and detector, data collection times may range from as little as 45 minutes to several days or even weeks.

8.5.5 Data processing (Chapter 5)

When the data collection stage is complete, the data are integrated and scaled and various background and absorption corrections are made.

8.5.6 Obtain .*hkl* file (Chapter 5)

After the data are processed, an .*hkl* file containing all the information about the reflections is written.

> The extensions of computer file names used in this text, such as .*hkl*, .*res*, and .*ins*, refer to those used by the SHELX suite of programs.

8.5.7 Determining the space group (Chapters 3 and 4)

Based on the .*hkl* file, an .*ins* (instruction) file is written with the unit-cell dimensions and the information on the molecular content of the crystal (the element type and number of atoms). With this information, combined with diffraction data, the space group can be determined.

8.5.8 Structure solution (Chapter 6)

After the space group determination step, attempts are made to solve the structure. The appropriate structure solution technique is selected and the relevant software is run. This process produces a .*res* (results) file that contains information on the unit cell, molecular content, and, if the structure solution process has been successful, lists the atoms their coordinates, occupancies, and isotropic thermal parameters.

8.5.9 Refinement (Chapter 7)

With the results of the structure solution in hand, the next step is to refine the structure. This is an iterative process; minor changes are made to the model, which is constantly compared to the electron density map. This process uses both the *.ins* and the *.res* file. These two files are, in essence, almost identical, with the *.ins* containing instructions to run for the refinement process while the *.res* file contains the results post-refinement.

After a cycle of refinement, the *.ins* file is overwritten with the *.res* file, producing, in effect, a 'new' *.ins* file to progress the refinement process further.

At the end of this process, once the refinement has reached convergence, the molecular model can then be analysed for interactions, such as hydrogen bonding or π-stacking.

The results of the final molecular model can also be used as models for computational calculations, such as molecular orbital or energy calculations.

8.5.10 Creating final files and publishing (Chapter 9)

With the convergence of the refinement process, the final step in the crystallographic process is the writing of the final files. In small-molecule crystallography, this is usually a *.cif* file, while for protein crystallography, this is usually either a *.pdb* or an *.mmCif* file.

Once these files are created and completed, they can then be validated and submitted for publication.

◉ CHAPTER SUMMARY

1. While modern diffractometers have enabled the crystal structure determination process to become fairly automated and routine, the challenge of growing good quality crystals suitable for diffraction remains.

2. The techniques most commonly used to grow crystals include:

    (a) Slow evaporation;

    (b) Solvent layering;

    (c) Vapour diffusion;

    (d) The hanging-drop technique (mainly for protein crystals).

3. Single crystals can be mounted on the diffractometer using a variety of techniques, such as mounting on a glass fibre, in a glass capillary or in a loop.

4. An X-ray diffractometer typically consists of a detector, an X-ray source, a collimator, and a goniometer to hold the crystal sample.

5. Laboratory X-ray sources are usually sealed tubes or rotating anodes, while synchrotron radiation is able to provide very strong and (optionally) tunable X-rays.

6. There are two main types of area detector, the charge-coupled device (CCD) and the image plate (IP).

7. The crystallographic process begins first with a visual examination of crystals leading to the selection of a 'good' crystal.

8. Once on the diffractometer, data are collected from the crystal structure, processed, and corrected.

9. An *.hkl* file containing all the diffraction data, together with *.ins* and *.res* files, form the core of the structure solution and crystallographic refinement process.

10. The final files from the crystallographic process are usually submitted to a database as part of the publication process.

⊜ REFERENCES

1. Clegg, W. (2000). Synchrotron chemical crystallography. *Journal of the Chemical Society: Dalton Transactions* 3223–32.

📖 FURTHER READING

Clegg, W., Blake, A. J., Gould, R. O., and Main, P. (2001). *Crystal Structure Analysis – Principles and Practice.* Oxford University Press/International Union of Crystallography, Oxford.

Drenth, J. (1999). *Principles of Protein X-ray Crystallography,* 2nd edn. Springer Advanced Texts in Chemistry. Springer-Verlag, New York.

Glusker, J. P., Lewis, M., and Rossi, M. (1994). *Crystal Structure Analysis for Chemists and Biologists.* VCH, New York.

Stout, G. H. and Jensen, L. H. (1989). *X-Ray Structure Determination: A Practical Guide.* John Wiley & Sons, Ltd, New York.

🕸 LINKS

Commonly miscible solvents: http://webpub.allegheny.edu/employee/s/smurphre/ research/miscibility.pdf

PUBLISHING CRYSTAL DATA 9

By the end of this chapter you should be able to:

- Identify and describe the parameters typically used to validate a crystal structure;
- Gain an appreciation of how one might share crystallographic information;
- Identify the main crystallographic databases;
- Understand the purpose and functions of crystallographic databases.

9.1 INTRODUCTION

Science relies on publishing (in journals) as a way of sharing the exciting and new information gained from months and often years of experimentation in the laboratory. This is equally true of the publication of crystal structures.

Results from the crystal structures of small molecules are often published as part of a larger overall paper – discussing the synthesis, analysis, and reactions of these small molecules, with the crystal structure being a form of identification of the new synthetic molecule. Protein crystal structures are often of interest in their own right as they frequently give us insight into their form and function.

There are also some specialist crystallography journals completely devoted to the science of crystallography, ranging from the development of new hardware and computer programs to discussions of the structure and the solid-state interactions within to the various experiments that push the boundaries of crystallography.

While crystal structures form an important contribution to the body of science, not all of the diffraction data collected or crystal structures that are solved and refined are valid for publication. Whether or not a crystal structure is 'good enough' to be published relies on certain standards.

With this in mind, it is important and necessary to be able to quantify what is deemed a 'good' structure solution. While we often aim towards a publishable standard of a structure solution, there may be times when it is sufficient for us simply to be able to identify the contents of a crystal lattice (perhaps as a form of confirmation). In both cases, we need to know how to validate the available information.

9.2 VALIDATION OF CRYSTAL DATA

Section learning outcomes

To be able to:

- Identify the main parameters used to validate a crystal structure;
- Understand and describe how some of the parameters may be used to ensure the end of the refinement process;
- Discuss how the parameters can contribute to the overall quality of the crystal structure.

As the crystallographic process draws to a close with the convergence of the refinement process, we then need to be able to identify the parameters and values that are typically used to validate a crystal structure solution.

The simplest of these to identify is the *R*-factor (Section 7.2.1) which is able to provide a numerical value to determine the 'goodness' of a structure. Other parameters that can contribute to the validation process include the GOOF, the resolution of data, the data-to-parameter ratio, and the completeness of data.

9.2.1 The *R*-factor or *R*-value

During the process of refinement, the *R*-factor or *R*-value is often used as a monitor of the progress of the refinement cycles. It is also often an overall indicator of whether the refinements have reached convergence.

$$R_1 = \frac{\sum \left\| F^{calc} \right| - \left| F^{obs} \right\|}{\sum \left| F^{obs} \right|}$$

At the start of the structure solution process, if a dataset is fitted with a model that has randomly placed atoms, the overall *R*-factor is likely to be very high. These values are estimated to be 0.83 for a centrosymmetric structure and 0.59 for a non-centrosymmetric structure.

In small-molecule crystallography, for a structure that is solved, one would expect *R*-values to be in the region of 0.25–0.15. Further refinement to convergence would usually bring this value closer to the range of 0.08–0.05 or less, although crystal structures with higher *R*-values have been published.

In macromolecular crystallography, it is often more common to have final *R*-values that are higher, as the molecules under study are often significantly larger with correspondingly larger unit cells and often with a higher degree of disorder within the structures. A typical final R-value would be in the region of 0.20, although if the structure is relatively small and ordered, it may sometimes be possible to obtain R-values of about 0.10.

While R-values are always written as decimals, they are almost always referred to in terms of percentages. For example, a crystal structure that is reported to have an R-value of 0.08 is said to be refined to 8%.

As well as the standard R-value, a weighted R-value is also usually quoted, although its value is often slightly higher that that of the standard R.

The weighted R-value is given as:

$$wR_1 = \sqrt{\frac{\sum w(F_o^2 - F_c^2)^2}{\sum w(F_o^2)^2}}.$$

The wR is calculated based on the square of the structure factors, F^2, where each reflection is weighted according to a standard weighting scheme. The purpose of the weighting scheme is to take into consideration all the random errors that may occur in the course of the diffraction experiment.

9.2.2 Goodness-of-fit (GOOF)

The goodness-of-fit, (GOOF) is another statistical measure, often assumed to provide an indication as to whether a structure is 'good'. The GOOF of a structure, when refined to convergence, is often a numerical value close to 1.

From a crystallographic point of view, however, the GOOF is a statistically weighted tool and can and often is induced to fit a value close to 1. This means that if the GOOF is a value that is quite distant from the required 1, the weighting scheme related to the GOOF can be changed prior to the final cycles of refinement, in order to make the value of the GOOF close to 1.

Overall, R-values provide a better overall view of the convergence of refinement and are more useful in helping the crystallographer determine whether or not a model is sufficiently accurate.

9.2.3 Resolution of data

Another factor that can influence the quality of the structural model is the resolution of the diffraction data. What is resolution? In essence, the resolution determines how well or how clearly we can differentiate two adjacent spots.

If we look at Fig. 9.1, we find that the two adjacent spots at (a) appear distinct and clear while the two adjacent spots at (b) are blurred and fuzzy. Generally we would say that (a) has good resolution and (b) is of poor resolution; or in comparison we could say that (a) has a higher resolution than (b).

The resolution of diffraction data from a crystal can be similarly defined in terms of high or low resolution. It is often influenced by the scattering power of the atoms within the crystal lattice; that is, how well the atoms are able to scatter X-rays. This, in turn, can be

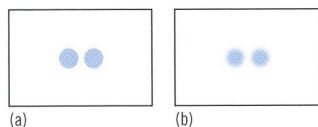

FIGURE 9.1 Two spots at (a) good resolution and (b) poor resolution

influenced by a variety of factors, from the temperature of the data collection (determining the thermal vibrations of the atoms) to the size of the crystal under study, to the set-up, type, and size of the detector on the diffractometer.

Overall, the resolution of diffraction data is defined in terms of angstroms (Å). Generally, this can be interpreted to mean that if the resolution of the data is collected to 2.0 Å, this broadly means that we should be able to differentiate two separate atoms at a distance of 2.0 Å.

The resolution of a diffraction dataset can be calculated based on Bragg's law, $d = \lambda / 2\sin\theta$. Experimentally, diffraction data from macromolecules are usually collected with copper (Cu Kα) radiation, which has a wavelength of 1.54 Å. Based on the equation, this would mean that one can theoretically expect resolution up to 0.7 Å, although practically it is often usual the find the resolution closer to 2.0 Å.

Diffraction data for small molecules are usually collected using the molybdenum (Mo Kα) wavelength of 0.71 Å. Again, theoretically this should equate to a resolution of approximately 0.35 Å but the typical experimental resolution for small molecules is about 0.8 Å.

It usually follows that the higher the resolution of data (the smaller the number) that can be collected, the better the quality of the structure solution and refinement.

9.2.4 Data-to-parameter ratio

Another factor that contributes to how 'good' a structure is solved and refined is the data-to-parameter ratio. The parameters that are usually refined through the refinement process (Section 7.2.2) are the atomic positions (fractional coordinates), all the atomic displacement parameters (both isotropic and anisotropic), and, if necessary, the occupancy (if less than unity).

The data-to-parameter ratio is the ratio of the number of diffraction spots or reflections obtained (during the experimental data collection) to the number of parameters that need to be refined through the refinement process.

The higher the data-to-parameter ratio, the more likelihood there is that the structure will refine well to a low R-value, and consequently we are more likely to be more confident of the results of the experiment.

Often a balance needs to be found between having sufficient data to fully resolve the structure and sufficient data to refine the parameters so as to achieve low e.s.d.s on atomic positions and bond parameters, both of which will provide a good indication of the quality of the crystal structure.

9.2.5 Completeness of data

Completeness of data refers to the percentage of data collected relating to the reciprocal lattice space. In essence, this relates to how much diffraction data, or how many unique reflections, need to be collected to be able to obtain a complete picture of the reciprocal lattice space.

Completeness of data is usually measured in terms of percentage and is defined as a percentage of the reciprocal lattice space.

If a dataset is 97% complete, this usually means that 97% of all the unique reflections have been collected with 3% missing. Of course this means that the better completeness of data that can be obtained, the more accurate the structure solution and refinements will be.

However, it is not always necessary to obtain 100% of the reciprocal lattice space, often it is possible to solve and refine small molecules to a decent level with 95% completeness of data.

SELF-TEST QUESTIONS

1. How might we determine whether a crystal structure has reached the end of the refinement process?

2. Describe how resolution is determined in crystallographic terms. Use an example to illustrate your answer.

3. Completeness of data and data-to-parameter ratios are two measures of how 'good' a crystal structure may be. How do you think one may influence the other? Explain your answer.

9.3 DEPOSITION OF FILES

Section learning outcomes

To be able to:

- Identify the relevant databases for the different types of crystal structures;
- Understand how crystal structures may be deposited in databases;
- Discuss why these databases are useful.

So, what happens after a crystal structure has been solved and refined and structurally analysed? Science, scientists, and even the general population benefit directly from shared information and this applies equally to crystal structures.

Apart from being routinely published in journals, crystal structures are collected in two main databases, the Cambridge Structural Database (CSD) for small molecules and the Protein Databank (PDB) for macromolecular structures. The PDB is accessible for free at http://www.rcsb.org, while the CSD is a commercially available database.

At the end of the crystallographic refinement process, a final *.cif* file (for small molecules) and the *.mmCif* or *.pdb* file (for macromolecules) is written. Within each of these files a host of information is included, for instance:

Details of the scientists or authors of the related paper;

Details of the crystal growing methods;

Experimental details can also be input, such as:

    Data collection set-up (details);

Type of diffractometer used;

Software used for data processing;

Types and use of corrections to data;

Details of data processing.

Details of the crystal structure:

Description of the crystal (e.g., size, colour, and shape);

Bravais lattice type;

Space group;

Cell dimensions and angles;

Cell volume and density.

Atomic details:

Atomic positions;

Atomic displacement parameters;

Occupancy.

Molecular details:

Bond lengths and angles;

Connectivity tables.

Details on refinements and final values.

When a structure is ready to be published, the *.cif* or *.pdb* file is routinely deposited with the relevant database at the same time. Although the databases do not routinely screen the structures, there are online crystallographic validation tools, which publishing authors are encouraged to use.

However, it is often up to the reader to be able to read critically and discern the quality of the published crystal structure. It is also possible to deposit structures with these databases prior to or without publication.

Databases serve as a 'one-stop' collection of all published crystal structures. Their computer interfaces are often equipped with several forms of search functions from structural searches via line molecular drawings to searches based on a molecular fragment, space groups, unit-cell dimensions, and a range of other possibilities.

9.3.1 The Cambridge Structural Database (CSD)

The Cambridge Structural Database (CSD), developed and maintained by the Cambridge Crystallographic Data Centre (CCDC, website: http//www.ccdc.cam.ac.uk), is a database of small-molecule crystal structures (of organic and organometallic compounds), containing bibliographic, chemical, and crystallographic information.

The crystal structures submitted to the database are based on X-ray or neutron diffraction studies and consist of crystal structures that have either been deposited as part of a journal publication submission or been submitted directly to the CSD as 'private communications'.

As of the 1st January, 2009, the CSD contains 469 611 structures and more are being continually added. Figure 9.2 shows how the number of structures contributed to the CSD has grown exponentially since 1972.

9.3.2 Protein Data Bank (PDB)

The Protein Databank (PDB)[1] is where the final files from the crystallographic process of macromolecules are deposited. Essentially it consists of submitted three-dimensional structures from large biological molecules (proteins and nucleic acids). These structures are determined from X-ray diffraction, NMR, and electron microscopy.

From the crystallographic process, *.pdb* or *.mmCif* files are the common file types that are submitted to the PDB. As of the 14th April, 2009, the PDB contained 57 013 structures; of these 48 859 were structures determined by X-ray crystallography.

As with the rapid growth of the number of structures in the CSD, the PDB (see Fig. 9.3) too has grown exponentially over time. Its numbers, however, remain significantly smaller than those in the CSD, mainly because of the relative ease with which small-molecule crystals can be grown and studied compared with the large macromolecules.

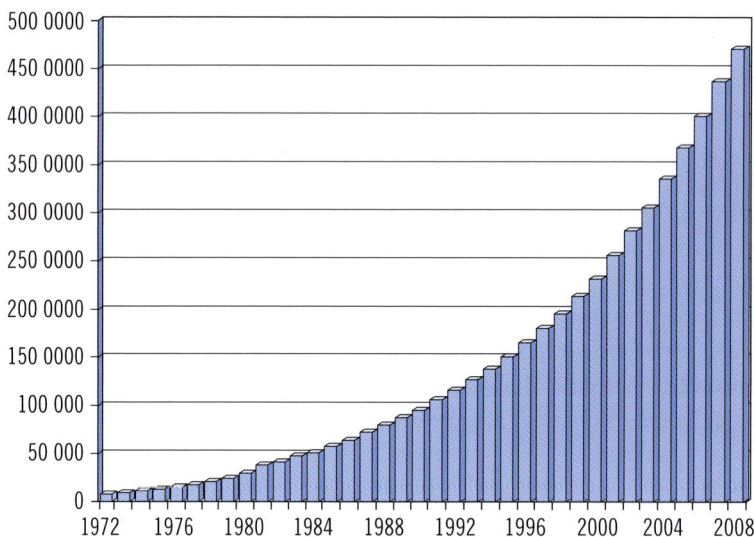

FIGURE 9.2 Number of crystal structures in the CSD since 1972

Reproduced with permission from: Allen, F. H. (2002). The Cambridge Structural Database: a quarter of a million crystal structures and rising. *Acta Crystallographica B* **B58**:380–8. DOI: 10.1107/S0108768102003890.

[1] RCSB Protein Data Bank (PDB) http//www.pdb.org

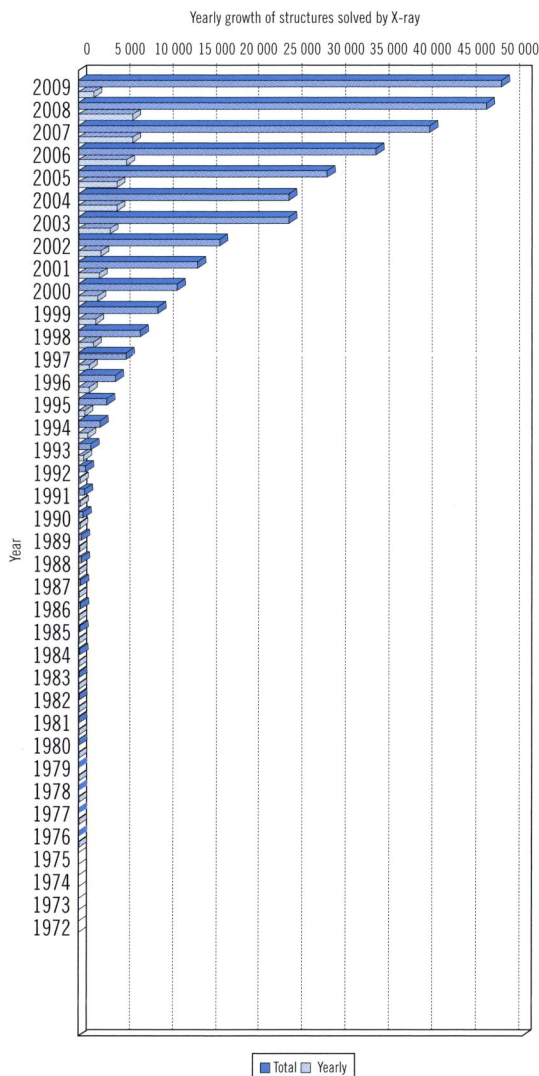

FIGURE 9.3 The number of X-ray determined crystal structures in the PDB since 1976

Reproduced with permission from: Berman, H. M., Westbrook, J., Feng, Z., *et al*. The Protein Data Bank. *Nucleic Acids Research* **28**:235–42 (2000). (http://www.pdb.org).

9.4 CONCLUSION

As the determination of crystal structures by X-ray crystallography is now routine, particularly with the development and advancement of hardware (X-ray diffractometers) and software (the related crystallographic computer software), it is natural then for the boundaries of crystallography to continue to be pushed and expanded.

While the challenge to be able to determine the smaller and more weakly diffracting crystals better will always remain, a host of other experiments in crystallography continue to excite and inspire.

Among these experiments are time-resolved crystallography, gas and liquid diffusion experiments into crystals, detailed charge density studies, and high pressure crystallography (subjecting crystals to extremely high pressure), which all seek to provide an insight into the exciting and dynamic processes and changes that may occur in the crystalline solid state.

◉ CHAPTER SUMMARY

1. Several different parameters can be used to validate crystal data. Among them are:
   (a) *R*-values;
   (b) Resolution of data;
   (c) Data-to-parameter ratio;
   (d) Completeness of data.

2. The consideration of *R*-values is different for macromolecules and small molecules. Standard *R*-values are typically higher for macromolecules.

3. The resolution of data influences the quality of the structural model.

4. Resolution, in crystallographic terms, may be defined as the measure of how well one is able to differentiate two adjacent spots.

5. The data-to-parameter ratios are determined by the number of reflections to the number of parameters that need to be refined. Typically, the higher the ratio, the better the quality of the structural model.

6. The completeness of data is the percentage of unique reflections collected from the reciprocal lattice space. This is again another measure of the quality of the diffraction data.

7. Once the refinement process is complete, the final files of the crystal structure are usually written (*.cif*, *.mmCif*, or *pdb*).

8. These files contain all relevant information pertaining to the crystal structure and the crystallographic process. Other details may also be included here.

9. The final files of the crystallographic process are usually submitted to a database upon publication.

10. The small-molecule crystal structure database is the Cambridge Structural Database (CSD) and macromolecular crystal structures are stored in the Protein Data Bank (PDB).

◉ FURTHER READING

Drenth, J. (1999). *Principles of Protein X-ray Crystallography*, 2nd edn. Springer Advanced Texts in Chemistry. Springer-Verlag, New York.

Giacovazzo, C., Monaco, H. L., Artioli, G., *et al.* (eds.) (2002). *Fundamentals of Crystallography*, 2nd edn. IUCr Texts on Crystallography. Oxford University Press, New York.

Glusker, J. P., Lewis, M., and Rossi, M. (1994). *Crystal Structure Analysis for Chemists and Biologists*. VCH, New York.

Stout, G. H. and Jensen, L. H. (1989). *X-Ray Structure Determination*: *A Practical Guide*. John Wiley & Sons Ltd, New York.

Wlodawer, A., Minor, W., Dauter, Z., and Jaskolski, M. (2008) Protein crystallography for non-crystallographers, or how to get the best (but not more) from published macromolecular structures. *FEBS Journal* **275**:1–21. (An excellent outline of how to judge published protein crystal structures.)

ANSWERS TO SELF-TEST QUESTIONS

Section 1.1

1. Some of the techniques that the chemist might employ are nuclear magnetic resonance (NMR), ultraviolet-visible spectroscopy, and infrared spectroscopy.

2. Spectroscopic methods make use of the vibrations within a molecule to allow inferences to be made regarding the molecular structure under study. Spectroscopy is able to provide information about the energy levels within a molecular system, which can then be used to infer details on molecular connectivity. In comparison, an X-ray diffraction experiment allows the direct determination of the molecular components within a crystal lattice, from which details on molecular connectivity can be obtained. X-ray crystallography requires that all samples are crystalline solids, while spectroscopic techniques can be used to analyse a range of sample types, from solids to liquids. X-ray crystallography can provide very detailed information about a crystal lattice but it is not necessarily a bulk technique, while spectroscopy can often be used to analyse the bulk of a sample.

3. Nucleic acids (derivatives of DNA and RNA) and globular proteins are most suitable for X-ray diffraction. It is possible to crystallize globular proteins as the large number of intramolecular and intermolecular interactions contribute to its crystallinity. The crystallization process of globular proteins is also often aided by the solvent molecules of the crystallization solution, which fills the spaces within the protein, often through hydrogen bonding.

Section 1.2

1. Solids can be divided into (a) crystalline and (b) amorphous solids.

2.
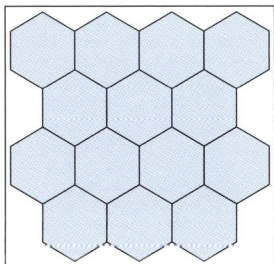 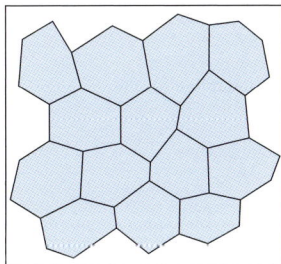

(a) (b)

The molecules within crystalline solids (a) are arranged in an ordered fashion, with specific distances between specific points/locations and this is repeated throughout the crystal lattice, while amorphous solids (b) lack this long-range order.

3. An example of a solid with short-range order is glass. Both types of solids with short-range order and long-range order have molecules that are *not* randomly placed. However, the molecular arrangement within long-range order is very uniform and this uniformity is repeated throughout the crystal lattice, while this is not found in solids with short-range order.

Section 1.3

1. Possible lattice points and lattice planes:
in Fig. 1.6 (a);

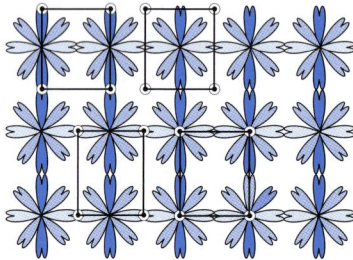

in Fig. 1.6 (b);

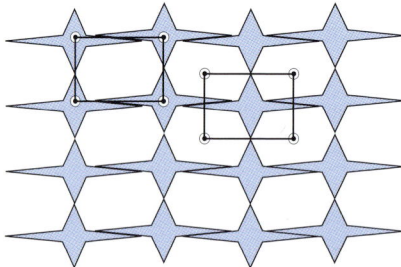

in Fig. 1.6 (c);

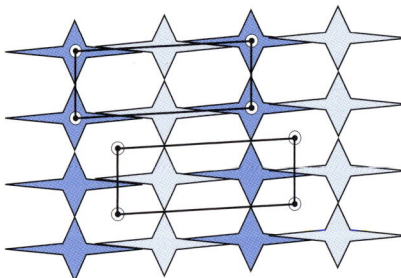

Section 1.5

1. The orthorhombic unit cell has all angles at 90° but all axes are not equal ($a \neq b \neq c$, $\alpha = \beta = \gamma = 90°$)

2. The unit cell parameters of a monoclinic cell are $a \neq b \neq c$, $\alpha = \gamma = 90°$; $\beta \neq 90°$

3. The unit cell in Fig. 1.12(a) clearly contains a dark blue and a light blue object; in combination this provides a unit cell with a higher symmetry than that of Fig. 1.12(b).

4. The *trigonal* unit cell can be derived from the hexagonal cell: a hexagonal cell can be subdivided into three trigonal cells (two whole cells and two half cells).

5. The face-centred lattice type (F) does not occur for the tetragonal crystal system because the unit cell can be reduced further to a body centred (I-type) tetragonal lattice.

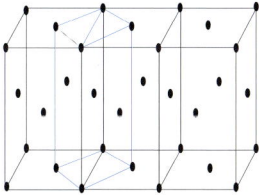

Section 2.2

Section 2.2.1

1. The Weiss indices are:

   (a) *a: b: c;*

   (b) *2a: 3b: 3c;*

   (c) *2a: 2b: 5c;*

   (d) *3a: 4b: 2c.*

2.

   (a) *2a: b: 2c;*

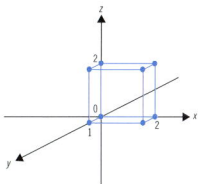

   (b) *3a: b: 5c;*

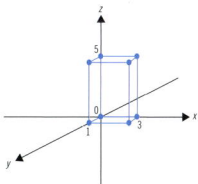

   (c) *a: 3b: 4c.*

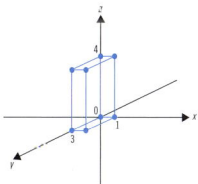

Section 2.2.4

1. The Miller indices are:

    (a) $(0, 0, 2)$;

    (b) $(1, 2, 0)$;

    (c) $(1, 0, 1)$;

    (d) $(1, 1, 0)$.

Section 2.3

1.

   (a) X-rays are suitable for single crystal diffraction experiments because the typical wavelengths for X-rays are of the same order as the molecular bond distances within a crystal lattice. This enables X-rays to interact with the contents of a crystal lattice, allowing a diffraction experiment to be carried out.

   (b) The most commonly used sources of X-rays are copper and molybdenum.

   (c) X-rays are produced from energy that is emitted when an electron in an excited state relaxes back to its ground state.

   (d) The radiation produced from electron movement depends on the atomic shells that are involved in the transition. When an electron moves from L to K, it emits $K\alpha$ (both type 1 and 2) radiation; when it goes from M to K it emits $K\beta$ radiation, and when it moves from M to L it emits $L\alpha$ radiation.

2. (Referring to Fig. 2.5) A diffraction pattern is produced when during an X-ray diffraction experiment, a crystal is irradiated with X-rays. The interaction between the oscillating electrons within each atom and the incoming X-ray beam causes the X-rays to diffract in all directions. The diffraction from all the Miller sets gives rise to '**diffraction spots**' which represent the **diffraction pattern** of the crystal being examined.

3.

   (a)

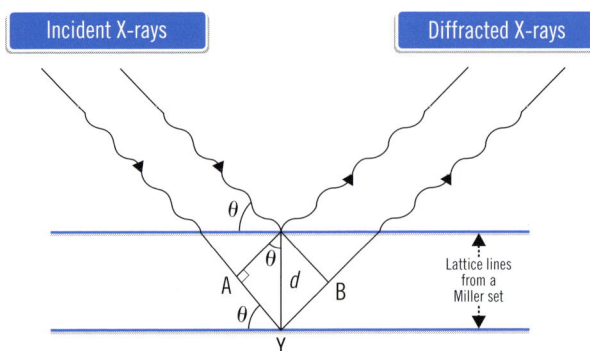

Within a crystal lattice, the lattice lines and planes (Miller sets) are located at a separation distance of d (this is also known as d-spacing). When the incident X-rays collide with the crystal lattice at specific points on the crystal lattice, the X-rays are reflected or diffracted out from the crystal lattice.

The path taken by the incident and reflected beam can be calculated as the sum of AY and YB and this is equal to the wavelength of the incident X-ray beam, $n\lambda$. If the angle at which the incident beam collides with the crystal is θ, then we can calculate the path difference in terms of d, θ, and λ.

(b) If the path difference is

$$AY + YB = n\lambda,$$

and we know that

$$\sin\theta = AY/d;$$

rearranging the equation,

$$AY = d\sin\theta \quad (AY = YB),$$

therefore, the path difference:

$$2d\sin\theta = n\lambda.$$

Section 2.4

1. Bragg's law states that $2d\sin\theta = n\lambda$; this equation can be rearranged to form

$$\sin\theta = \left(\frac{n\lambda}{2}\right)\left(\frac{1}{d}\right).$$

The inverse relationship between $\sin\theta$ and d proposes that when a lattice has a short axis, that is, small values of d, the spacing of the diffraction spots would appear further apart.

2. Real space is related to reciprocal space through the inverse relationship of $\sin\theta = (n\lambda/2)(1/d)$. This relationship can also be represented pictorially by:

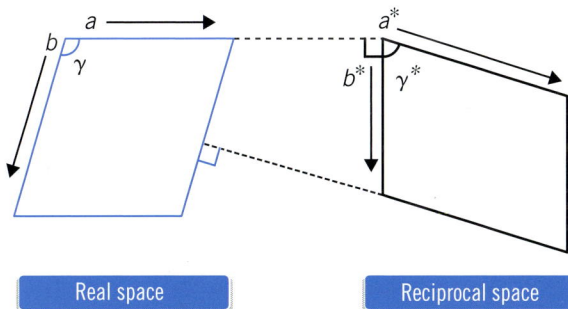

Real space   Reciprocal space

A mathematical correlation between real and reciprocal space can be given by the following equation. Assuming that the starred values are those of the reciprocal then:
$a.a^* = b.b^* = c.c^* = 1; \alpha = \alpha^*; \beta = \beta^*; \gamma = \gamma^*.$

Section 3.2

1. Translational symmetry incorporates only a translation of an object, while non-translational symmetry would contain a form of rotation. Translational symmetry causes absences in crystal data, while non-translational symmetry does not cause absences.

2. Translational symmetry elements are screw axes and glide planes while non-translational symmetry elements are the inversion centre, rotation, reflection, and rotary-inversion.

3. The translation of an object in a 2_1 screw axis consists of a rotation of $(360/2) = 180°$, then translation upwards by half the unit cell axis. In the diagram below, object 1 is rotated 180° then translated up by half a unit cell axis to its position at 2. The same object is then rotated by 180° and translated upwards again by half the length of a unit cell to its original position in the next unit cell.

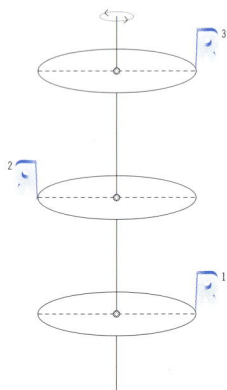

Section 3.3

1. The types of symmetry operation that can occur in a lattice type are also dependent on the shape of the lattice. For that reason, not all of the types of symmetry operator can arise in all of the 14 Bravais lattice types. For example, it is impossible for a glide plane to occur in a triclinic lattice that has no equivalent axes and no faces that lie at right angles.

2. The only type of symmetry element that may occur in a triclinic lattice is an inversion centre.

3. A screw axis cannot occur in a triclinic lattice because the shape of the lattice, which has unequal axes and no angles at 90°, will not allow the translation of objects within the triclinic lattice.

4. Within an orthorhombic lattice, there should be at least three two-fold rotations.

Section 4.2

1.

   (a) $P\,2\,2\,2$ denotes a primitive lattice with two-fold rotations along each of the three axes;

   (b) $C\,2$ denotes a C-centred lattice with a two-fold rotation along the b-axis;

   (c) $C\,m$ denotes a C-centred lattice with a mirror plane perpendicular to the b-axis;

   (d) $I\,b\,a\,2$ denotes a body-centred lattice with a b-glide perpendicular to the a-axis, an a glide perpendicular to the b-axis and a two-fold rotation along the c-axis.

2.

   (a) a glide perpendicular to b;

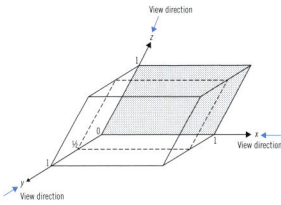

(b) *b* glide perpendicular to *c*;

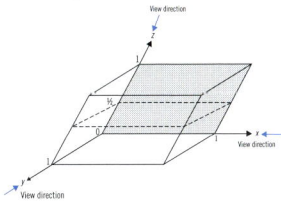

(c) *b* glide perpendicular to *a*.

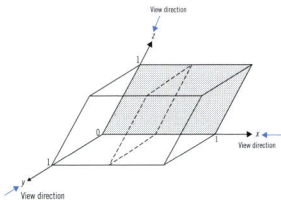

3. Based on the diagram, an object located at (x, y, z) is translated along the y-axis by half the length of the unit cell and then reflected across the glide plane to its final position at $(-x, y + \frac{1}{2}, z)$.

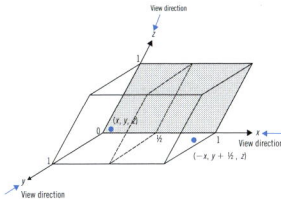

Section 4.3

1. $P\,2_1\,2_1\,2$ contains screw axes along the x- and y-axes and a two-fold rotation along the z-axis, while $P\,2_1\,2_1\,2_1$ has screw axes along each of the three axes.

2. The equivalent positions in the $P\,2_1/m$ space group are (x, y, z), $(-x, \frac{1}{2} + y, -z)$, $(x, \frac{1}{2} - y, z)$, and $(-x, -y, -z)$. The other Pt atoms are likely to be at $(-0.3, 0.8, 0.6)$, $(0.3, 0.2, -0.6)$, and $(-0.3, -0.3, 0.6)$.

3. The equivalent position diagram for orthorhombic space group $P\,2_1 2_1 2_1$ is:

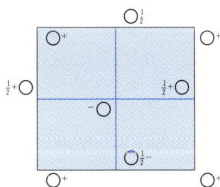

Section 4.6

1. An atom is considered to be in a special position when its symmetry-equivalent point maps onto itself by another symmetry operation.

2. Centrosymmetric space groups contain a centre of symmetry (usually an inversion centre), while non-centrosymmetric space groups do not contain a centre of symmetry. An example of a centrosymmetric space group is the triclinic $P\bar{1}$, while a non-centrosymmetric space group is $P2_12_12_1$.

3. Sohnke space groups describe chiral space groups in which only chiral molecules can crystallize. All proteins are chiral molecules, as their amino acid building blocks are chiral. As a result, proteins can only crystallize in one of the 65 chiral space groups. An example of a chiral molecule is L-guanine.

Section 5.2

1. An *absence* occurs on a diffraction map when the value of $I_{hkl}^{obs} = 0$. This means that there is no discernable diffraction at that particular point in the diffraction map.

2. When the complete diffraction data are collected, the data are processed. This stage is known as *data reduction*, in which diffraction data from each Miller plane are integrated across the multiple frames of diffraction data. After integration, scaling and various corrections for background take place, and, finally, the information is written in an *.hkl* file.

3. At low Bragg angles, the atomic scattering factors are directly proportional to the atomic numbers of the elements; however, this tails off at higher Bragg angles.

4. The reduction of atomic scattering factors at high Bragg angles tails off more slowly for heavy elements than for light atoms, causing the heavy elements to dominate a diffraction map.

5.

1	2	1	0.91	0.82
2	2	0	−4.20	17.65
2	0	2	−4.73	22.38
0	2	2	−1.15	1.32

Section 5.3

1.

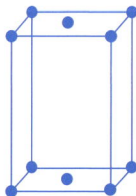

For a C-type Bravais lattice, assuming that each lattice point is equivalent to a single atom of type J, the coordinates of atoms at the eight corners of the unit cell are

$(0, 0, 0)$, $(1, 0, 0)$, $(0,1,0)$, $(0,0,1)$, $(1,1,0)$, $(0,1,1)$, $(1,0,1)$, and $(1,1,1)$. Each atom makes a 1/8 contribution to the unit cell.

As the lattice points (in this case atoms of type J) at the corners of the unit cell are all equivalent, any one of these can be selected to count as the total contribution from all the corners of the unit cell.

Let us select the atom at $(0, 0, 0)$.

The atoms at the centres of the two 'C' faces of the unit cell have fractional coordinates ($\frac{1}{2}$, $\frac{1}{2}$, 0) and ($\frac{1}{2}$, $\frac{1}{2}$, 1). Each of these atoms make a contribution of $\frac{1}{2}$ to the unit cell. As these atoms are also equivalent, either of these can be selected as the total contribution from the corners of the unit cell. Let us select the atom at ($\frac{1}{2}$, $\frac{1}{2}$, 0).

Substituting into Eqn 5.2, we get an expression for the general structure factor for this Bravais lattice:

$$F_{hkl} = f_J \cos 2\pi(0) + f_J \cos 2\pi\left(\frac{1}{2}h + \frac{1}{2}k\right).$$

Simplifying the equation:

$$F_{hkl} = f_J + f_J \cos\pi(h+k).$$

Keep in mind that when an absence occurs, the intensity, I, is equal to zero ($I=0$). (Remember also that $I=F^2$.)

From the equation, it becomes clear that the structure factors will have a value of zero when $\cos\pi(h+k)=-1$. This will only occur when $h+k$ is an odd number. Thus, a C-type Bravais lattice will be identified when all reflections in a dataset for which $h+k$ is an odd number are absent, i.e., have no intensity. This is often written as $h+k=2n+1$.

2. The general absences for:

    (a) B-type Bravais lattice: $h+l=2n+1$ (odd and absent);

    (b) F-type Bravais lattice: reflections must have either all even or all odd indices to be observed. Mixed odd and even indices are not allowed;

    (c) I-type Bravais lattice: $h+k+l=2n+1$ (i.e., sum of indices odd).

3. (a) No general absences denote a primitive (P) lattice.

    (b) Systematic absences corresponding to:

    $(0, k, l)$, where values of k odd are absent: b glide perpendicular to a;

    $(h, 0, l)$, where values of l odd are absent: c glide perpendicular to b;

    $(h, k, 0)$, where values of h odd are absent: a glide perpendicular to c.

    The space group is orthorhombic P bca.

4. No general absences.

    Systematic absences corresponding to:

    $(h, 0, 0)$: h odd absent;

    $(0, k, 0)$: k odd absent;

    $(0, 0, l)$: l odd absent.

Section 6.2

1. To model a crystal structure, it is necessary to know approximately the number and types of atoms to be found within the crystal lattice. It is also necessary to know the amplitude, direction, and phase of the incident X-rays.

2. The phase problem refers to the loss of the information of the incident phase angle when the incident X-rays are diffracted from a crystal lattice. As a crystal lattice consists of a repeating array this is akin to a periodic function and subsequently the phase problem can be described in terms of the Fourier transform.

3. Real space can be related to reciprocal space with the Fourier transform as shown in the table below:

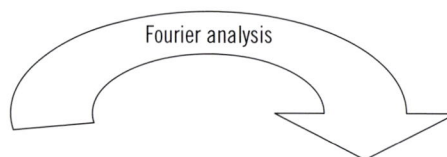

Fourier analysis

Real space	Reciprocal space
Crystal structure	Diffraction pattern
Electron density (atomic parameters)	Structure factors, Amplitudes and phases (X-rays)
Crystal lattice; unit cell	Reciprocal lattice; cell
Coordinates (x, y, z)	Coordinates (h, k, l)

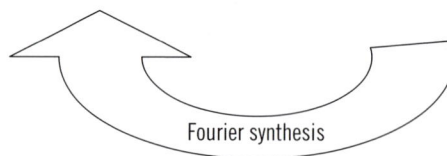

Fourier synthesis

Section 6.3

1.

(a) Crystallographers may use direct methods to solve the structure of the organic ligand as it does not contain heavy metals; they may use Patterson methods to solve the structure of the Hg complex, which contains a heavy element and is more suited to the Patterson method of structure solution.

(b) Direct methods are based on guessing or estimating the possible phase angles and must satisfy two important conditions, which are:

(i) That the value of electron density (ρ) must never be negative; it may be positive or zero but never negative; and

(ii) That the electron density maps have to have high values (sharp peaks) at and near atomic positions and must have nearly zero values everywhere else.

The Patterson method uses a trial structure to calculate the relative phase angles. This is done by taking the squared values of the structure factors and this removes the need for the phase angle information from the Fourier equation.

2. The similarities between both methods of structure solution are:

    (a) Both are methods in structure solution for single-crystal X-ray diffraction;

    (b) Both are only used once in the process of structure solution;

    (c) Both are used to overcome or attempt to overcome the phase problem.

3. The differences in the two methods are detailed in the table below:

	Direct methods	Patterson methods
1.	Used mainly for organic or light-atom structures.	Used only for organometallic or structures containing heavy atoms.
2.	Based on selecting most important reflections and working out the probable phase relationships.	Based on calculating F_{hkl}^2 rather than F_{hkl} with all phases set to zero.
3.	The electron density map often contains recognizable features of the molecule under study.	The Patterson map shows vectors between the heavy atoms in the structure.
4.	Assumptions: Electron density is zero or close to zero where there are no atoms; Electron density is never negative.	Peaks in the Patterson map show where atoms lie relative to each other.

Section 6.4

1. Direct methods, isomorphous replacement (IR), and multi-wavelength anomalous diffraction (MAD).

2. The conventional direct-method approach is similar to that for small molecules, in which two conditions need to be fulfilled:

    (a) The electron density must never be negative;

    (b) In electron density maps, areas near atomic positions have high values while the electron density is zero everywhere else.

    The shake-and-bake method, however, uses a significant amount of computing power in attempting to model random arrangements of atoms, simulating each of the diffraction patterns, and subsequently comparing the simulated pattern with that of the actual recorded pattern.

3. IR is isomorphous replacement. The technique of IR uses the phases of a known solved protein structure as an initial model or as phase estimates for the new similar protein, where the phases of the known protein are used in the new set of experimental data as a phase model for the new protein.

4. Anomalous scattering refers to the inequality in the incident and diffracted intensity of the X-rays

Section 7.2

1. Refinement is a process that takes place after a structural model has been obtained.

    (a) The parameters that are refined are the atomic positions and the atomic displacement parameters.

(b) These parameters are continually moved within the iterative refinement process so that the overall energy minimum is found.

(c) The refinement process can be monitored by observing the R-value or the R-factor. The refinement process is usually complete when the R-value reaches a global energy minimum and is the lowest possible constant value over the course of refinement.

2. The least squares method is a statistical tool to compare with a specific value of certainty that a solution, or model, is agreeable to the experimentally obtained data.

   In crystallography, this means that the least squares method can provide numerical evidence as to how well the structural model (obtained from the structure solution process) fits the experimental diffraction data.

   The least squares method is based on the assumption that the experimental errors follow a normal distribution.

3. Errors can be quantified by using estimated standard deviations (e.s.d.). These are usually stated as a value of σ, written as one significant figure after a refined parameter. Crystallographers estimate errors to 3σ.

4. Data collection at low temperatures can help to maintain crystallinity and reduce thermal vibrations in the atoms and this subsequently enables more accurate structure solutions and smaller anisotropic vibrations within atoms from better and more intense diffraction data.

   Low temperatures can usually be achieved with the use of a cryostat that passes a stream of cold air to envelope the crystal during data collection. The use of liquid nitrogen lowers temperatures to approximately 100 K while helium lowers the temperatures to the order of 10 K.

Section 7.3

1. Disorder occurs when some of the atoms in the structure adopt different orientations in different unit cells in the crystal. When disorder occurs, the model in the solved structure represents two or more orientations of the atoms that are observed.

2. Disorder can be broadly divided into two categories: static disorder and vibrational or dynamic disorder.

   Static disorder can occur when there are two conformers of the same molecule within the same unit cell or the conformers face in one direction in some unit cells and in another direction in other unit cells, while vibrational or dynamic disorder occurs when the molecule is constantly undergoing flux. A typical example is that of ferrocene, which undergoes a phase transition at 164 K, where close to this temperature the ring orientations of the ferrocene molecule are in constant motion

3. Merohedral twinning occurs when two crystal lattices are connected in such a way that from the collected diffraction data, they appear to be single crystal. The diffraction data of the twinned crystal consequently exhibit a higher symmetry than should be the case. Non-merohedral twinning, however, is usually easier to recognize as the diffraction data are not easily resolved and there are clear signs of more than one reciprocal lattice, due to overlapping spots in the diffraction pattern.

4. Polymorphism occurs when a molecule is known to exist in more than one crystalline form.

   The most well known form of polymorphism is the polymorphism of carbon, which produces graphite or diamonds. While both are carbon-based, the physical character-istics of each are very different.

Section 7.4

1. Both hydrogen bonding and π–π interactions are interactions that can be observed in the solid crystalline state. Hydrogen bonds are non-covalent interactions that specifically involve donor – acceptor interactions, where the donor is a coordinated hydrogen atom, while π–π interactions occur between two (or more) adjacent aro-matic phenyl rings.

2. The three types of hydrogen bonds that can occur are strong, moderate, and weak hydrogen bonds.

   Strong hydrogen bonds are also known as 'ionic hydrogen bonds', are often formed in groups where there is a deficiency of electron density in the donor group, O^+–H, N^+–H, or an excess of electron density in the acceptor group, F^-, O^-–H, O–C, O–P or N^-.

   Moderate hydrogen bonds are mostly formed by neutral donor and acceptor groups, such as O–H, N–H, and O=C, where donor atoms are electronegative relative to the hydrogen and the acceptor, B, has lone-pair unshared electrons.

   Weak hydrogen bonds are formed when the hydrogen atom is covalently bonded to a slightly more electroneutral atom relative to hydrogen, for example, C–H, Si–H. Weak hydrogen bonding also occurs when the acceptor group has no lone pairs but has π electrons, such as C≡C or an aromatic system.

3. π–π interactions generally involve an interaction between two or more phenyl rings.

4. The types of geometry that can be found for π–π interactions are the face-to-face geometry, the edge-on or T-shaped geometries, and the offset π-stacked geometry.

5. The four types of stacking pattern that can be found for π–π interactions are: (a) the herringbone motif; (b) the sandwich herringbone; (c) the γ-motif, and (d) the β-motif.

Section 8.3

1. Two techniques commonly used in growing crystals are slow evaporation and vapour diffusion (other techniques include layering and the hanging-drop technique).

   Slow evaporation occurs when a saturated solution of the compound is left to evapo-rate slowly. The evaporation of the solvent causes supersaturation to occur and subse-quently crystals may form.

   Vapour diffusion occurs when the compound is dissolved in a small, open container and the small container is placed in a larger container with a volatile but miscible non-solvent. As the non solvent diffuses into the solution, supersaturation occurs and crystals may form.

2. A good quality crystal, one that has clear faces and clean edges, usually has a higher possibility of producing good quality diffraction data. The structure solution and refinement processes that follow are subsequently more successful with the use of a good quality crystal.

3. Air-sensitive crystals may be mounted on a glass fibre coated in inert perfluoroether oil or mounted in a sealed glass capillary.

Section 8.4

1. An X-ray diffractometer consists of an X-ray source, a detector, a collimator to pare down the X-ray beam and a goniometer to mount the crystal.

2. The typical sources of X-rays for diffractometers are based on X-ray generators. These are: (a) the sealed X-ray tube, (b) the rotating anode, and (c) the synchrotron.

   The sealed tube and the rotating anode are sources of X-ray generators for laboratories. A rotating anode is usually capable of producing more intense X-rays than the sealed tube. The use of synchrotron radiation however, offers very-high-energy X-rays, which are often suitable for crystals whose structures cannot be determined on a laboratory diffractometer.

3. The most common type of modern detector are (a) the charged coupled device (CCD) and (b) the image plate (IP).

   On an X-ray diffractometer, the detector is located after the crystal, where it is able to 'collect information' from the diffracted X-rays.

   The CCD consists of an array of pixels (these can be thought of as electron buckets that can be filled and emptied). The CCD is made of a layer of a type of semiconductor material (usually silicon), which is then covered with a transparent metal oxide film.

   Energy from the diffracted X-rays can liberate electrons from the surface of the CCD, which then creates holes that are positive. These holes are then repelled by the already applied positive potential and move to the base, while the electrons are collected in a potential well. The stronger the source of X-rays, the more electrons are collected in the wells. The charge is then moved along channels (or rows), by means of varying potentials, towards the data processing device. This is a very fast process that occurs in fractions of seconds, resulting in a readout of a 'frame' of diffraction spots.

   An image plate is a very flexible image sensor that consists of a polyester support film, coated with a layer of microcrystals of luminescent material (for example $BaFBr:Eu^{2+}$).

   When the diffracted X-rays reach the image plate, the electrons in the microcrystals are excited and then become trapped in the lattice of the empty anion (F^-). This causes the colour of the luminescent material to change. A laser is used to scan the IP and the information is channelled and then transferred to a photomultiplier tube, where the information is converted into electric signals, which are then processed and digitized to produce a 'frame' of diffraction spots.

Section 9.2

1. The refinement process is near the end when the *R*-values remain consistently low throughout the final cycles of refinement.

2. Resolution in crystallographic terms describes how well two atoms can be separated and is usually defined in terms of angstroms (Å). For example, if the data are collected at a resolution of 1.8 Å, this broadly means that two separate atoms at a distance of 1.8 Å can be differentiated.

3. Completeness of data refers to the percentage of data collected relating to the reciprocal lattice space, while the data-to-parameter ratio is the ratio of the number of diffraction spots or reflections obtained (during the experimental data collection) to the number of parameters that need to be refined through the refinement process. Usually it follows that a high completeness of data increases the data-to-parameter ratio. 'Completeness of data' refers to the amount of diffraction data that are collected. A 'high completeness of data' value means that a larger number of data points is available to be refined against for each parameter.

INDEX

Page references in bold indicate tables and those in italic indicate figures.